WALCH  PUBLISHING

Write

Daily Warm-Ups

CHEMISTRY

Brian Pressley

Level II

1   2   3   4   5   6   7   8   9   10

ISBN 0-8251-5062-0

Copyright © 2004

Walch Publishing

P. O. Box 658 • Portland, Maine 04104-0658

walch.com

Printed in the United States of America

The *Daily Warm-Ups* series is a wonderful way to turn extra classroom minutes into valuable learning time. The 180 quick activities—one for each day of the school year—review, practice, and teach chemistry topics. These daily activities may be used at the very beginning of class to get students into learning mode, near the end of class to make good educational use of that transitional time, in the middle of class to shift gears between lessons—or whenever else you have minutes that now go unused. In addition to providing students with fascinating chemistry activities, they are a natural path to other classroom activities involving critical thinking.

Daily Warm-Ups are easy-to-use reproducibles—simply photocopy the day's activity and distribute it. Or make a transparency of the activity and project it on the board. You may want to use the activities for extra-credit points or as a check on critical-thinking skills and problem-solving skills.

However you choose to use them, *Daily Warm-Ups* are a convenient and useful supplement to your regular lesson plans. Make every minute of your class time count!

Intro to Chemistry

The world around you is made up of trillions of particles that are too small to see. These bits of matter follow rules and laws that allow us to identify them and to predict how they will interact with one another. It has often been said that chemistry is the study of matter, its properties, and its behavior. Because we are surrounded by matter, chemistry is one of the most important sciences in that it allows us to observe and then *change* our environment. The question remains, however: How much should we change our environment, and can we predict the results before we do so?

Name either 10 items that you could find in a typical household that are a direct result of people studying chemistry, or 10 ways that chemistry has changed the life of the average person, for better or for worse.

1

All That Matters

There are many ways to classify the materials we see around us every day. Is it a solid, liquid, gas, or plasma? Is it an element or a compound? Is it a pure substance or a mixture? The answers to these questions allow us to identify the things we see. For example, we learn to tell the difference between an apple and a banana, or either of those from a ham sandwich. All of these items are food, they all have nutrients, like vitamins and minerals, but they are not identical. Without the ability to observe the environment and identify what we are seeing, we would be unable to respond to and use the materials around us.

2

Name some properties that you would use to tell the difference between a banana and a ham sandwich. Try to think of properties that could be used to tell *any* banana from *any* ham sandwich.

Unique Properties

We need to be able to identify the materials we see every day so that we can use them correctly. We do this by observing physical and chemical properties and comparing them to what we know from personal experience. Physical properties include such things as boiling point, color, density, hardness, melting point, odor, taste, and even electrical conductivity. Chemical properties are usually a measure of how a material reacts or fails to react with other substances. Will it burn? Does it dissolve in water? Does it produce bubbles of gas when dropped into acid? All of these things allow us to tell the difference between water and alcohol, for example, and many other substances with the use of only one or two of our senses.

List at least five of the physical and chemical properties of water and alcohol. What simple test could you do to determine which liquid was which?

Measuring Up

In the 1790s, a group of scientists came to the first general agreement that using a single, worldwide system of measure would benefit all people. This system was named the metric system, and it gave us units of measure like meters and liters. The system was formally, and more universally, adopted in the 1960s when it became the International System of Units, or SI system. The "SI" comes from the French name Le Systèm International d'Unités. For example, the SI scale of temperature is called Celsius, and it places the freezing point of water at 0°C and the boiling point of water at 100°C.

What are some of the reasons why defining a temperature scale by the freezing and boiling points of water might be useful to scientists on any part of the earth?

4

Significant Uncertainty

If you were to weigh a small rock on a scale that could measure the mass of the rock to the nearest 0.001 grams, then the mass of the rock would be, for example, 10.871 ± 0.001 grams. The last digit is really just the best estimate of what the last digit should be. Perhaps it was rounded or perhaps not—there is no way to be certain—so the last digit is called *uncertain*. The first four digits were numbers about which no estimate was made, so they are called *significant figures*. All nonzero numbers are significant; zeros between nonzeros are significant; place-holding zeros at the beginning and end of a number are not significant; and zeros at the end of a number after the decimal are significant.

How many significant figures are in each of the following?

a. 8.01

b. 80.1

c. 80

d. 8009

e. 0.0083

f. 0.1040900300

5

From Another Dimension

Just as there are 12 inches in one foot, there are 100 centimeters in one meter. The labels on the ends of these numbers are called units or dimensions. *Dimensional analysis* is the process of changing the units on a number, usually to make a number more manageable. We might say a certain event took place 20 years ago, but we would seldom say that it took place 7,300 days ago. The amount of time is the same, but by changing the years into days, we change not just the unit, but the number in front of the unit.

Convert the following.

a. 600 days to years

b. 14 centuries to years

c. $3.50 to quarters

d. 900 centimeters to meters

The Theory That Matters

John Dalton, an English schoolteacher, proposed the atomic theory of matter at the beginning of the 1800s. It states

1. Each element is made of small particles called atoms.

2. All atoms of one kind of element are identical.

3. Atoms can't be created or destroyed by chemical reaction or changed into other kinds of atoms.

4. Compounds are formed by combining atoms of more than one kind of element, in the same ratio each time.

What, if any, problems arise from the basic ideas of Dalton's theory?

7

Building Blocks of the Building Blocks

Not every discovery in science is the direct result of a scientist looking for a specific item and then finding it. There are plenty of examples in which scientists involved in research in one area are surprised to find that they have made a major advance in another. For example, take the three major subatomic particles: the electron, the proton, and the neutron. The proton and the neutron are roughly the same size, and the electron is over 1,800 times smaller than either the proton or the neutron. These particles were all discovered in the course of research into the nature of the atom, but none of the scientists who found them were actually looking for subatomic particles as such. However, in all three instances the new information led to an explosion of new thinking and ideas about the nature of matter.

How can proper experimental methods prepare a researcher for the event of making an unexpected discovery?

8

Isotopes

The atomic mass of each element is given on the periodic table and is actually a weighted average of all the naturally occurring isotopes of an element. An isotope is a type of atom that has the same number of protons but a different number of neutrons. For example, all carbon atoms have six protons, but may have six, seven, or eight neutrons, so there are isotopes of carbon-12, carbon-13, and carbon-14. Because carbon-13 and carbon-14 occur in much smaller amounts than carbon-12, they don't change the atomic mass of carbon very much, but it is still 12.0111 u and not exactly 12 u.

If fluorine has an atomic mass of 18.998 and there are isotopes of both fluorine-18 and fluorine-19, which of the two isotopes occurs in larger numbers naturally?

9

Mass and Weight

In chemistry it is important to distinguish between mass and weight. Mass is a measure of the amount of matter in an object, while weight is the gravitational force that pulls on that amount of matter. It is necessary to make measurements in science in terms of mass, because mass has an unchanging standard of measurement for a sample of matter, while weight may fluctuate from location to location. This would apply during a chemical reaction or to a measurement made on the equator, the moon, or in orbit around the earth, where the effects of gravity are nullified by free fall.

Name three items people buy that are measured in mass and not weight. Why is this an advantage to the consumer?

10

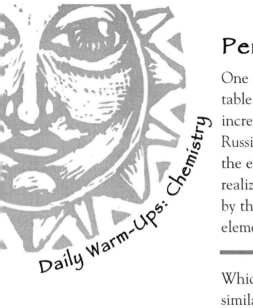

Periodicity

One of the many ways scientists classify substances is by the periodic table. The periodic table is a chart showing the elements in order by increasing atomic number and grouped by their similar qualities. Russian chemist Dimitri Mendeleev is believed to be the first to place the elements in order by their increasing atomic masses. Mendeleev realized that the repeating patterns he saw were grouping the elements by their properties, and he even went so far as to leave spaces for elements that had not yet been discovered.

Which two of the following elements should show the most similarities: C, Se, Xe, Si, Ag, and Mg?

11

Molecular and Empirical

A *molecular formula* is a representation of all the elements contained in a molecule and the number of atoms of each element in that molecule. For example, the substance sulfuric acid has the molecular formula H_2SO_4. An *empirical formula* is the simplest ratio of the elements in a formula. So for benzene, C_6H_6 is the molecular formula, but the ratio of carbon to hydrogen is 6:6, or 1:1, so the empirical formula is CH.

Write empirical formulas for the following.

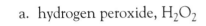

    a. hydrogen peroxide, H_2O_2

    b. glucose, $C_6H_{12}O_6$

    c. water, H_2O

    d. ethane, C_2H_6

12

Ions

An ion is an atom or a group of atoms that have acquired an electrical charge due to a loss or gain of electrons. The atoms of some elements in the periodic table tend to gain or lose one or more electrons in an effort to have an outer electron shell like that of the closest noble gas. They do this to become more stable, which, in a way, is a goal of all elements that are free to react. Column 1 of the periodic table usually produces ions with a +1 charge; Column 2 usually produces +2 charges; and the transition metals in Groups 3–12 usually have a +1 or +2 charge. Column 13 often has +3 ions, and Columns 15, 16, and 17 usually produce ions with charges of –3, –2, and –1, respectively. The noble gases don't readily form ions and almost always have a charge of zero.

What is the most likely ionic charge that each of the following elements would form?

a. Li   c. Br   e. P

b. Ne   d. Ca   f. Ga

13

What's in a Name?

A binary compound is a simple sort of compound usually made up of ions. The compound made when a positive and negative ion combine has a collective charge of zero and is not an ion as a whole. Table salt made up of sodium and chlorine is an example of this. The sodium ion has a +1 charge, and the chloride ion has a charge of −1. The salt itself has the formula NaCl and has no collective charge. In naming such compounds, the name of the positive ion comes first and is the same as the name of the element. The name of the negative ion comes second and is changed to have an -ide ending. For example, oxygen becomes oxide, fluorine becomes fluoride, and bromium becomes bromide. The number of each element does not affect the name of the compound.

Name the following binary compounds.

    a. $BaCl_2$ c. Li_2O e. Na_2S

    b. KBr   d. MgO   f. Be_3N_2

Covalent Names

Binary molecular compounds have a simple naming system. These are typically the result of a bond forming between two nonmetals, such as sulfur and oxygen. The number of each element present determines the name of the compound, a system based on common prefixes. Each element gets a prefix that shows how many of each element is available, except when there is only one atom of the first element. The prefix mono- is never used for the first element. For example, NO_2 is nitrogen dioxide, S_2O_3 is disulfur trioxide, and P_5O_{10} is pentaphosphorus decoxide.

Number of atoms	1	2	3	4	5	6	7	8	9	10
Prefix	mono	di	tri	tetra	penta	hexa	hepta	octa	nona	deca

Name the following.

a. N_2O_2

b. CCl_4

c. PBr_6

d. NO

e. S_4O_9

f. C_5I_7

15

Equation for Success

A chemical equation is a shorthand statement that uses chemical formulas to describe a chemical reaction. From the burning of gasoline in a car motor to the digestion of an apple, whenever a chemical reaction takes place, it must have certain qualities that make it work. There must be reactants, the chemicals that initially come into contact; and there must be products, the new substances that are formed when atoms combine chemically with other atoms.

$$2H_2 \quad + \quad O_2 \quad \rightarrow \quad 2H_2O$$

This simple equation describes the combination of oxygen and hydrogen to form water. The coefficients in front of the hydrogen and water formulas make the equation balanced, and they are there to express the amount of each reactant and product needed to make the equation true.

Balance the following equations.

a. $C_2H_2 \quad + \quad O_2 \quad \rightarrow \quad CO_2 \quad + \quad H_2O$

b. $NaCl \quad \rightarrow \quad Cl_2 \quad + \quad Na$

c. $Ti \quad + \quad N_2 \quad \rightarrow \quad Ti_3N_4$

16

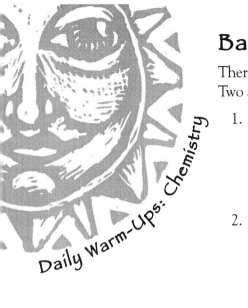

Basic Reaction Types

There are some kinds of reactions that chemists see over and over again. Two such types are the combination and decomposition reactions.

1. *Combination reaction* (synthesis)—a reaction in which two or more elements combine to form a single substance.

$$A \ + \ B \ \rightarrow \ AB$$

$$2H_2 \ + \ O_2 \ \rightarrow \ 2H_2O$$

2. *Decomposition reaction* (analysis)—a reaction in which a single substance separates into two substances.

$$AB \ \rightarrow \ A \ + \ B$$

$$2H_2O \ \rightarrow \ 2H_2 \ + \ O_2$$

Determine which of the following are combination and which are decomposition reactions.

a. $Al \ + \ Cl_2 \ \rightarrow \ Al_2Cl_3$

b. $K_2CO_3 \ \rightarrow \ K_2O \ + \ CO_2$

c. $SO_2 \ + \ O_2 \ \rightarrow \ SO_3$

d. $KClO_3 \ \rightarrow \ KCl \ + \ O_2$

17

Formula Mass

Formula mass is the sum of all the atomic masses in a compound. For example, NaCl, or sodium chloride, is made of sodium and chlorine. The atomic mass of sodium is 22.990 u, and the atomic mass of chlorine is 35.453 u. So the combined masses of 22.990 u + 35.453 u = 58.443 u, which is the formula mass of sodium chloride. If there were subscripts in the formula of the substance, the element with the subscript would have to be counted the same number of times as the number of the subscript. For example, in K_2O, the potassium is counted twice and the oxygen once, so: 39.098 u + 39.098 u + 16.000 u = 94.196 u.

Daily Warm-Ups: Chemistry

Find the formula mass of each of the following.

a. KCl

b. $BaCO_3$

c. P_4O_{10}

d. C_6H_5OH

e. CaO

f. $KClO_4$

18

The Mighty Mole

In chemistry the *mole* is a way of keeping track of a large group of particles and also a way of determining the amount of matter in a substance. The mole can be thought of in much the same way as the dozen. A dozen eggs is 12 eggs, a dozen cars is 12 cars, and a dozen of any kind of object is 12 of that object. A mole is a number equal to 6.02×10^{23} particles. So a mole of pickles would have 6.02×10^{23} pickles in it. To give you some idea of how big a mole is, one scientist estimated that a mole of peas would cover the entire surface of the earth, oceans and all, to a depth of 6 inches. As you can see, the mole is not a unit we would use in counting everyday objects; however, it is well suited for counting atoms and molecules because there may be trillions and trillions of them in a very small sample.

Answer the following.

a. How many moles of atoms are in 4.04×10^{24} atoms?

b. How many atoms are in 3.20 moles of atoms?

c. How many moles of carbon atoms are in 1.75×10^{25} carbon atoms?

19

Combustion

A hydrocarbon *combustion reaction* is a reaction in which a hydrocarbon is burned and consumed in the presence of oxygen. This type of reaction goes on all around us, from propane stoves and butane lighters to the octane in gasoline and the wax in paraffin candles. In the example below, C_xH_y is representative of any possible hydrocarbon—e.g., CH_4 or C_2H_6.

$$C_xH_y \; + \; O_2 \; \rightarrow \; CO_2 \; + \; H_2O$$

$$2C_2H_6 \; + \; 7O_2 \; \rightarrow \; 4CO_2 \; + \; 6H_2O$$

Balance the following combustion reactions.

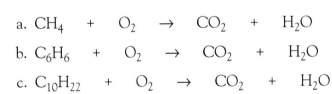

a. $CH_4 \; + \; O_2 \; \rightarrow \; CO_2 \; + \; H_2O$

b. $C_6H_6 \; + \; O_2 \; \rightarrow \; CO_2 \; + \; H_2O$

c. $C_{10}H_{22} \; + \; O_2 \; \rightarrow \; CO_2 \; + \; H_2O$

Daily Warm-Ups: Chemistry

20

Stoichiometry

Stoichiometry is the quantitative use of chemical equations and formulas to predict the amounts of products consumed or reactants produced during a chemical reaction. For example, the reaction of hydrogen and oxygen to form water,

$$2H_2 \quad + \quad O_2 \quad \rightarrow \quad 2H_2O$$

shows us that the ratio of hydrogen to water is 2:2. For every two molecules of water that are formed, two molecules of hydrogen must be consumed. Likewise, for every single molecule of oxygen consumed, two molecules of water would be produced. In both cases, we assume there are enough reactants for the entire reaction to take place.

How many molecules of hydrogen and oxygen would be needed to make 40 molecules of water?

21

© 2004 Walch Publishing

The Sky's Not the Limit

At some point during a chemical reaction, the reactants run out, and the reaction stops. This happens when either all the reactants run out at the same time or, more likely, when one of the reactants runs out first. In this case, the element that runs out first is called a *limiting reactant*. For example, if you were making cheese sandwiches and you ran out of bread before you ran out of cheese, the bread would be like a limiting reactant.

Given the formula $2H_2 + O_2 \rightarrow 2H_2O$, if you had 45 molecules of hydrogen and 45 molecules of oxygen, which of the two would be the limiting reactant?

22

Pass the Electrolytes, Please

An *electrolyte* is a substance whose aqueous solution will conduct electricity. In other words, if it is dissolved in water the resulting solution will carry an electric current. This is important to people—we have a mixture of electrolytes in our bodies, and the cells in our bodies use them to regulate the electric charge and flow of water molecules across our cell membranes. This affects our endurance, mental state, strength, and reflexes and the efficiency of our nervous system. This is why many popular sports drinks boast that they replenish the electrolytes that athletes lose when they sweat.

If your nervous system uses electricity to send commands to various parts of the body from the brain, then why is it so important that we maintain a proper level of electrolytes at all times?

23

Double Up

A *double-replacement reaction* is a reaction in which two ionic compounds exchange ions to create two new ionic compounds. Such reactions often take place when the reactants are in the form of an aqueous solution. In some cases, one or more of the resulting products is insoluble in water and settles out as a solid called a precipitate. For example:

AB   +   CD   →   AD   +   CB

$AgNO_3$ + $NaCl$ → $AgCl$ + $NaNO_3$

Complete the following double-replacement reactions.

a. $BaCl_2$ + K_2SO_4 →   +

b. $NaOH$ + HCl →   +

c. NH_4Cl + $NaOH$ →   +

24

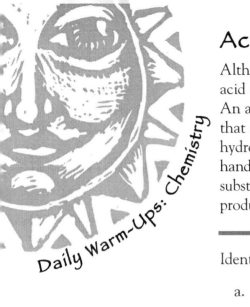

Acids and Bases

Although there is some disagreement on the exact definition of an acid or a base, there are some general qualities that help define them. An acid is a) a substance that can be a proton donor, b) a substance that can act as an electron acceptor, and c) a substance that produces hydronium ions (H_3O^{1+}) when dissolved in water. On the other hand, a base is a) a substance that can be a proton acceptor, b) a substance that can act as an electron donor, and c) a substance that produces hydroxyl ions (OH^{1-}) when dissolved in water.

Identify each of the following as an acid or a base.

    a. H_2SO_4 _____

    b. NaOH _____

    c. HCl _____

    d. $HClO_4$ _____

    e. KOH _____

    f. $Ca(OH)_2$ _____

25

Oxidation and Reduction

In many chemical reactions, the outermost electrons of an atom constitute the major contribution to the interactions between all the atoms in the reaction. Sometimes they move from one atom to another, allowing the atoms to bond or to become more stable. *Oxidation* is a reaction in which a particle—such as an ion, an atom, or a molecule—loses an electron. When this occurs, the oxidation number of some element in the reaction increases. *Reduction* is a reaction in which a particle—such as an ion, an atom, or a molecule—gains an electron, and the oxidation number of some element in the reaction decreases.

26

Identify the element in each reaction that is oxidized and each element that is reduced.

a. $2Mg + O_2 \rightarrow 2MgO$

b. $C + O_2 \rightarrow CO_2$

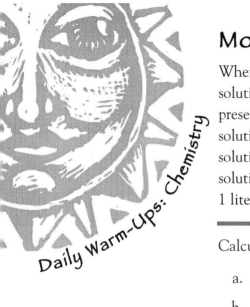

Molarity

When dealing with substances that have been dissolved into a solution, it is necessary to determine the amount of material that is present in a sample of solution. Molarity is the concentration of a solution given in moles of solute per liter of solution. A 1-molar solution of NaCl in water would have 1 mole of salt in 1 liter of solution. A 3-molar solution of NaCl would have 3 moles of salt in 1 liter of solution.

Calculate the molarities of the following solutions.

a. 3.4 moles of salt in 4.0 liters of solution

b. 5.0 moles of potassium chloride in 2.4 liters of solution

c. If you have a 6.0 molar salt solution with a volume of 2.0 liters, how many moles of salt are in it?

27

© 2004 Walch Publishing

Titration

Titration is the process by which the concentration of a solution is determined by reacting it with a solution of known concentration. For example, if two chemicals are known to react in a 1:1 ratio, then a solution of known concentration, often called a standard solution, can be made and reacted with a solution that has an unknown concentration. This process is usually carried out in the presence of a third chemical, called an *indicator*, that will change the color of the mixing solutions when one or the other of the chemicals has been completely used up or is in excess. If one milliliter of standard solution is used up in a reaction with one milliliter of unknown solution, then they have the same concentration. If twice as much standard solution is used, then the concentration of the unknown is double that of the standard solution.

28

How much standard solution would be used up in a titration in which the concentration of the unknown is five times greater than that of the standard? How about one half of the standard?

Time to Expand

Energy is the ability to do work, usually measured in joules. Energy is found in many forms and classifications, such as kinetic, potential, nuclear, electrical, chemical, thermal, mechanical, and sound. Scientists have long thought that matter is also a form of energy, and the law of conservation of energy states that the total amount of energy in the universe is constant, but it can change from one form to another. During a chemical reaction, the amount of matter that is present at the beginning of the reaction should be that same as the amount found at the end of the reaction, and the amount of energy should remain constant throughout the reaction.

How can sound be considered a form of energy?

The First Law

The first law of thermodynamics states that energy is conserved. In other words, the amount of energy in a system can change, but any energy lost during a chemical reaction must be gained by its surroundings. Although a fire may give off heat, the energy does not just disappear—it is spread throughout its surroundings until it is so spread out that it can no longer be measured by the human senses.

What is the change in internal energy of a system in a reaction in which 300 joules are absorbed by the system from its surroundings and 120 joules of that energy are used to do work on the surroundings?

30

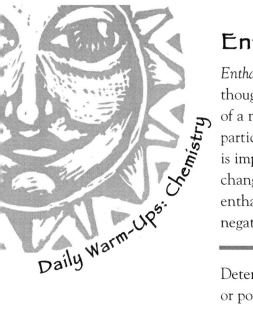

Enthalpy

Enthalpy is a measure of the internal energy of a system. It can be thought of as the sum total of all the energy in the chemical bonds of a material, the energy of its state, the kinetic energy of its internal particles, and any other way that a substance can contain energy. It is impossible to measure the enthalpy of a substance, but it does not change unless energy enters or leaves the substance. The value of enthalpy is positive when heat is absorbed by a substance and negative when given off from a substance.

Determine whether the enthalpy of each of the following is negative or positive.

a. water boiling

b. ice melting

c. gasoline burning

31

Enthalpy and Reactions

During chemical reactions, energy is usually gained or lost by some of the components in a reaction. Exothermic reactions give off heat while endothermic reactions absorb heat. The sum of those changes is the enthalpy of reaction, given by the formula

$$\Delta H = H_{products} - H_{reactants}$$

For example, in the reaction below, ΔH is negative, and therefore the reaction is exothermic.

$$2H_2(g) \quad + \quad O_2(g) \quad \rightarrow \quad 2H_2O(g) \qquad \Delta H = -483.6 \text{ kJ}$$

32

Tell if each of the following reactions is endothermic or exothermic.

a. $C(s) \quad + \quad O_2(g) \quad \rightarrow \quad CO_2(g) \qquad DH = -394 \text{ kJ}$

b. $N_2(g) \quad + \quad O_2(g) \quad \rightarrow \quad 2NO \qquad DH = 180.8 \text{ kJ}$

c. $S(s) \quad + \quad O_2(g) \quad \rightarrow \quad SO_2(g) \qquad DH = -297 \text{ kJ}$

Exploding Food

Calorimetry is a method for measuring heat flow into or out of an object. For example, we can measure the heat flow out of a potato chip by burning it rapidly (sometimes exploding it in the process) in a bomb calorimeter and then funneling the heat from the flame to a container of water. Then we can observe the temperature change of the water. Using specific heat, which is the measure of the amount of energy needed to raise the temperature of 1 gram of substance 1°C (1 K), we can determine the amount of heat that enters or leaves the material.

The specific heat of water is 4.19 J/g°C. How much heat is needed to raise the temperature of 18.0 grams of water 15°C?

Hess's Big Step

Germain Hess is known primarily for his law that states "the heat evolved or absorbed in a chemical process is the same whether the process takes place in one or in several steps." Also known as the law of constant heat summation, it is illustrated below.

Step 1 H_3PO_4 + $NaOH$ $\rightarrow$ NaH_2PO_4 + H_2O

Step 2 NaH_2PO_4 + $NaOH$ $\rightarrow$ Na_2HPO_4 + H_2O

Step 3 Na_2PO_4 + $NaOH$ $\rightarrow$ Na_3PO_4 + H_2O

If the energy from Step 1 is A, the energy from Step 2 is B, and the energy from Step 3 is C, then Hess's law says total energy = A + B + C.

34

For the reaction

$$H_3PO_4 \quad + \quad NaOH \quad \rightarrow \quad Na_3PO_4 \quad + \quad H_2O,$$

how does the energy gained or lost by this reaction compare to the energy gained or lost by the three equations above?

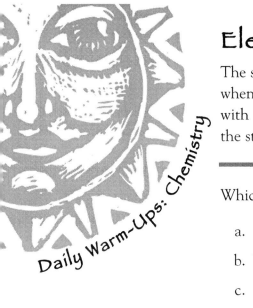

Elemental Enthalpy

The standard enthalpy of formation, ΔH°_f, is the change in enthalpy when a reaction forms one mole of a compound from its elements, with all the elements in their standard states and usually at 298 K as the standard temperature.

Which of the following elements are in their standard states?

a. mercury (solid)

b. hydrogen (gas)

c. carbon (solid)

d. nitrogen (liquid)

35

© 2004 Walch Publishing

Energy to Burn

We take many materials found on Earth and use them for fuel. This fuel does many jobs, from heating our houses to running our cars to generating electricity. Burning wood, for example, releases about 18 kJ of energy for every gram burned. Gasoline generates about 47 kJ/g when burned, and pure hydrogen can generate more than 140 kJ/g when burned. Each of these fuels has its advantages and disadvantages that make it useful in certain situations.

If hydrogen has the most energy to be released per gram, why don't we use it any time we need to burn a fuel to make energy?

36

A Constant Light

Light is a form of radiation that surrounds us almost all the time. Electromagnetic radiation takes on the form of radio waves, microwaves, infrared light, visible light, ultraviolet light, X rays, and gamma rays. All of these forms of energy have one thing in common: When traveling through a vacuum, they all move at the same speed, which is the speed of light. That speed is 300,000,000 meters per second, or 3.0×10^8 m/s.

Velocity of light is constant and equals the frequency of the wave times the wavelength. If blue light has a shorter wavelength than red light, how do the frequencies of the two compare?

37

Walking the Planck

German scientist Max Planck theorized that energy could only be released or absorbed from an atom in pieces of fixed size. He called these packets of energy quanta and said they were the smallest amount of energy that an atom could emit or absorb. He said that the energy, E, released or absorbed, had to equal the product of the frequency and a constant that is now called Planck's constant. The value of the constant is 6.63×10^{-34} Joule-seconds.

How much energy is released by a photon of blue light with a frequency of 6.00×10^{14} Hz?

38

A Spectrum of Models

Different substances absorb or emit light based on their atomic or molecular structure. Each element has a unique structure that is revealed through spectroscopy, which allows scientists to take a complex sample, expose it to heat, and look at the spectrum of emissions. The spectrum in its entirety looks like a rainbow, but the spectral lines of an individual element may consist of only four or five single lines at different points in the spectrum. The emissions can then be compared to emissions of known elements, and the elements that are present can be identified.

What kind of problem might arise if a person were trying to identify a large group of substances by looking at their spectra with just the naked eye?

39

Uncertain at Any Speed

German physicist Werner Heisenberg realized that it is impossible to know both the location and velocity of a subatomic particle at the same time. This concept, the *uncertainty principle*, is supported at the subatomic level, because when a photon collides with a subatomic particle, it can actually change the velocity or location of the particle. The photon comes back to the observer with perhaps a hint of where the particle was or how fast it was moving, but not both.

If particles of any size obeyed the same basic physics, why can we say with some degree of certainty that we know the location and velocity of a large object, such as a baseball or a car?

40

Orbitals

In the quantum mechanical model, an *orbital* describes an area of space in an atom where there is a high probability of finding an electron. The principal quantum number represents the number of different energy levels in an electron and is defined by whole counting numbers from 1 to n, where n represents whole counting numbers like 2, 3, and 4. There are almost always electrons in the levels below the highest energy level of an atom.

For an atom with a principal quantum number of 5, list all of the levels that would most likely contain electrons.

41

© 2004 Walch Publishing

s, p, d, f

The azimuthal quantum number denotes how many sublevels are in an atom, and it is represented by the letter l. The sublevels are further represented by the letters s, p, d, and f, and have 2, 6, 10, and 14 electrons as the maximum allowed in each level, respectively.

How many electrons would be in the fourth energy level of an atom if the 4s, 4p, 4d, and 4f orbitals were all full?

42

Electron Segregation

There are certain areas inside of every atom where electrons may be found. These areas, called orbitals, obey an idea called the *Pauli exclusion principle*. This principle says the only way two electrons can be in the same orbital is if they have opposite spins. Spin refers to the actual motion of the electron, which is not unlike the rotation of the earth on its axis. A box represents a single orbital, and arrows placed in the box represent the spin of the electron. Two electrons in an s orbital box would be represented as below.

What would the p orbital shown here look like with four electrons in it?

Electron Configurations

An *electron configuration* is the arrangement of electrons in the various orbitals of an atom. Electron configurations can be written using a form of notation that includes the principal energy level, the sublevel, and the number of electrons in that sublevel. For example, $3p^6$ means that this atom has electrons in the third principal energy level, in the p sublevel, and that there are six electrons. The general order that the orbitals fill for the ground state of all the elements is

$$1s^2\ 2s^2\ 2p^6\ 3s^2\ 3p^6\ 4s^2\ 3d^{10}\ 4p^6\ 5s^2\ 4d^{10}\ 5p^6\ 6s^2\ 4f^{14}\ 5d^{10}\ 6p^6\ 7s^2\ 5f^{14}\ 6d^{10}\ 7p^6$$

Write the electron configuration of mercury.

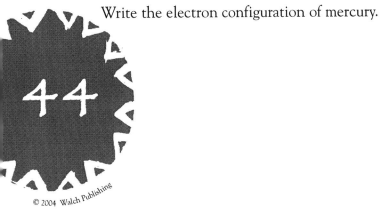

44

Periodic Design

The organization of elements in the periodic table lends itself well to a discussion of the orbitals found in atoms. The first two rows contain elements that have all their outermost electrons in their s orbitals. Columns 13–18 contain elements that have all their outermost electrons in their p orbitals. Columns 3–12 contain elements that have all their outermost electrons in their d orbitals. And the lanthanide and actinide series contain elements that have all their outermost electrons in their f orbitals.

Determine which orbital is the outermost for each of the following elements.

a. U

b. Sr

c. C

d. Ne

e. Fe

f. Pu

g. Na

h. Ag

45

Mendeleev and Masses

One of the many ways scientists classify substances is the periodic table. The periodic table is a chart showing the elements in order by increasing atomic number and grouped by similar qualities. In the late 1800s, Russian chemist Dimitri Mendeleev placed the elements in order by their increasing atomic masses and helped determine much about the order of the elements on the table. He left gaps for elements that had not been discovered at the time.

Why did it turn out to be better to place elements in order on the periodic table by atomic number instead of atomic mass?

46

Shielding

There is a shielding effect felt by the outermost electrons of an atom. This effect is the result of the interactions between the core electrons of an atom and the protons in the nucleus. You can think of the core electrons as soaking up some of the attractive force of the protons and not allowing their full attractive power to reach the valence electrons. One way to quantify this effect is to calculate the effective nuclear charge by subtracting the number of core electrons from the number of protons in the atom.

What are the effective nuclear charges of argon and calcium?

47

Atomic Size

There are two basic trends in atomic size on the periodic table.

1. In each column, the atomic size (radius) tends to increase as you move from the top to the bottom of the column.

2. In each row (period), the atomic size tends to decrease as you move from left to right.

Following the guidelines above, determine which of each pair would most likely be larger.

a. Mg or Ca

b. Y or Ru

c. C or Ge

d. Ne or Xe

e. W or Au

f. Zn or Hg

48

Energy to Escape

The *ionization energy* of an atom is the minimum amount of energy needed to remove an electron from an atom. The atom must be in its ground state, and must be in gaseous form. As each electron is removed, the ability of the protons in the nucleus to hold the remaining electrons increases. Consequently, as each subsequent electron is removed, it requires more and more energy to do so. Ionization energy generally increases from left to right in the rows and from bottom to top in the columns of the periodic table.

Place the following atoms in order by their increasing first ionization energy (the energy needed to remove just the first electron): Ar, He, Kr, Ne, Xe.

49

Electron Affinity

Electron affinity is a measure of the energy change that occurs when an atom gains an electron. The variation in electron affinity values varies widely across the rows of the periodic table. Many of the columns, on the other hand, show very little change in electron affinity when you move from the top to the bottom of a row. For example, fluorine, chlorine, and bromine have values of −328, −349, and −325, respectively.

Why might magnesium have a much higher electron affinity than sodium?

Metals

A *metal* is an element that loses electrons easily in a chemical change and has the properties of high luster, electrical and thermal conductivity, malleability, and ductility. Some examples of metals are sodium, chromium, and copper, and such metals are found toward the left and middle of the periodic table.

Since metals are elements that tend to lose electrons easily, what sort of ions are metals most likely to form?

51

Nonmetals

A *nonmetal* is an element that easily gains electrons during a chemical reaction and whose properties contrast with those of the metals. Some of the nonmetals are oxygen, nitrogen, and chlorine. Such nonmetals are found toward the top right of the periodic table.

Since nonmetals are elements that tend to gain electrons easily, what sort of ions are nonmetals most likely to form?

52

Hydrogen is Different

Hydrogen is placed above the alkali metals on the periodic table, but aside from having one valence electron, it has very little in common with the rest of the elements in that column. Hydrogen is a nonmetal, although it can appear metallic under high pressures, and it is a major component of water, fuels, acids, and almost all organic compounds. Hydrogen also forms both a positive and a negative ion.

What is one possible reason that hydrogen's properties are so unlike that of lithium, which is only one position below it on the periodic table?

53

© 2004 Walch Publishing

Eight Is Enough

In general, atoms tend to lose, gain, or share electrons until they are surrounded with eight valence electrons. This behavior allows them to emulate the stable arrangement usually found in the noble gas that is closest to them. This often results in their becoming much less chemically active, but only because in the process they often form bonds with other atoms for stability.

Tell whether each of the following would be more likely to gain or lose electrons to become stable.

a. Na

b. N

c. F

d. Li

e. Ar

f. S

54

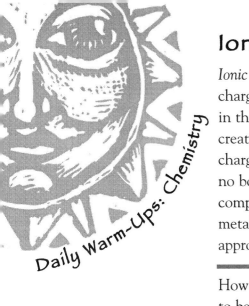

Ionic Bonds

Ionic bonds are formed by the electrical attraction between positively charged cations and negatively charged anions. The atoms involved in the bonding allow a transfer of electrons to take place, which creates a charge on each atom, making it an ion. The opposite charges then attract one another, forming a bond. Strictly speaking, no bond is completely ionic, because the electrons are not completely removed from the potential cation. But for most of the metals in the first two columns of the periodic table, the approximation is quite close.

How many electrons would each of the following have to gain or lose to become stable while forming an ionic bond?

a. Na

b. Mg

c. Al

d. Si

e. P

f. S

g. Cl

h. Ar

55

Covalent Bonds

In *covalent bonding,* the sharing of electrons gives bonds their strength. Atoms tend to form bonds to give them a more stable internal arrangement. Typically in covalent bonding, a nonmetal bonds to itself or another nonmetal through the sharing of electrons. Notice that chlorine has seven outer electrons and needs to have eight to be the most stable. If two chlorine atoms share one pair of electrons, both get to be surrounded by eight electrons.

How could oxygen form the molecule O_2 by covalent bonding if it only has six valence electrons?

56

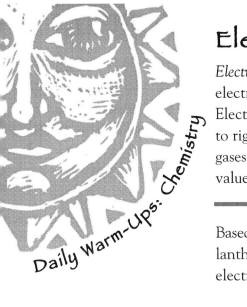

Electronegativity

Electronegativity is a measure of the ability of an atom to attract electrons to itself while it is chemically combined with another atom. Electronegativity generally increases in strength as you move from left to right on the periodic table and from bottom to top. The noble gases generally do not form compounds in nature, so electronegativity values are not usually calculated or listed.

Based on the general trends listed above, and discounting the lanthanides and actinides, which element should have the highest electronegativity? Which would have the lowest?

57

Lewis Dot Diagrams

Using Lewis symbols, sometimes called dot diagrams, is a way of writing elemental symbols with dots around them (and sometimes small circles or the letter x) to indicate the electrons that are free to be involved in bonding. Elements from Column 1 of the periodic table are usually shown with one dot; those in Column 2 get two dots; Column 13 gets three dots; 14 gets four; 15 gets five; 16 gets six; 17 gets seven; and Column 18, the noble gases, gets eight to show that the outer electron level is full.

Draw a dot diagram for each of the following elements.

58

a. Li

b. Be

c. B

d. C

e. N

f. O

g. F

h. Na

Resonance: Real and Wrong

Some molecules cannot be accounted for in terms of Lewis structures. It seems as though to make them workable, the octet rule would have to be violated. Take, for example, sulfur dioxide. There are only enough electrons to account for a double bond between sulfur and one of the two oxygens, which would leave the other oxygen with only a single bond. We know from experimentation that both bonds are the same, so we write two structures and show both possible situations, one where Oxygen 1 has the double bond and then one where Oxygen 2 has it.

Under what conditions should resonance be used to describe the bonds formed in a molecule?

59

When Eight Isn't Enough

In some cases there are molecules that simply have an odd number of valence electrons and do not follow the octet rule. When such a situation arises, chemists must go beyond the basic guidelines and come up with a new way to account for the location of all the electrons in the molecule. One of the most common compounds with this problem is nitric oxide (NO). Two such possibilities are given below.

:N: :Ö: :N: :Ö:

Draw a third possible configuration that NO could have.

60

Convalent Strength

Covalent molecules get their stability from the strength of their bonds. Chemists measure the strength of those bonds by breaking them. Bond enthalpy is a measure of how much energy is needed to break the bond between two atoms while they are combined in the gaseous phase. For example, it takes 163 kJ/mole to break the bond between two nitrogens, while it takes 567 kJ/mole to break the bond between hydrogen and fluorine.

What does the value of the bond enthalpy tell us about the stability of a bond?

61

Bond Angles

The shapes of various molecules are determined by their bond angles. A bond angle is the angle formed by the imaginary lines that join the nuclei of atoms combined in a molecule. Any two-atom molecule has a bond angle of 180°, which means it is essentially a straight line. A three-atom molecule can have a bond angle of 180°; otherwise, it must have a bond angle that is less than 180°, such as water, which has a bond angle of about 105°. As the bond angle gets smaller and smaller, the atoms of a molecule often become more and more crowded.

If the bond angle in an atom was very small, how might we expect that to affect the stability of the molecule?

62

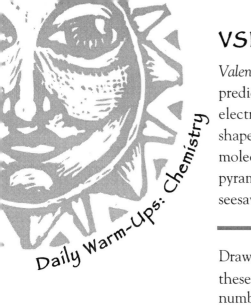

VSEPR

Valence-shell electron-pair repulsion (VSEPR) is a model that helps predict the shape of molecules. The way atoms and nonbonding electrons are attracted and repulsed by one another determines the shape of the molecule. There are many different shapes that simple molecules take on, such as linear, bent, trigonal planar, trigonal pyramidal, T-shaped, tetrahedral, octahedral, trigonal bipyramidal, seesaw, square planar, square pyramidal, and pentagonal bipyramidal.

Draw Lewis structures for methane, CH_4, and ammonia, NH_3. Both of these molecules are tetrahedral even though they have a different number of atoms. Why?

63

North and South?

Just as the earth and other magnetic bodies have a north and a south pole, some molecules, because of their bonds and shapes, have poles of a different kind. Because the molecules have positive and negative charges, the possibility that one end of a molecule will have more than its fair share of charge allows the molecule to have a positive and a negative end. Because each atom in a bond has a varying ability to attract charges to itself, some charges can get pulled to one side or the other within a molecule.

Why is HCl a polar molecule, but Cl_2 is not?

64

Hybridization

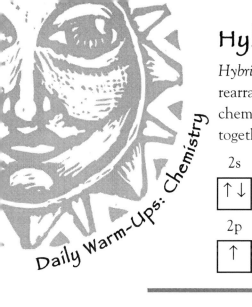

Hybridization is the process by which valence electrons in an atom rearrange themselves within the valence orbitals of an atom during a chemical reaction. For example, an s orbital and a p orbital can blend together to form an sp orbital. Thus,

2s

↑ ↓

becomes

sp³

↑ | ↑ | ↑ | ↑

2p

↑ | ↑ |

What would that hybridization of the boron atom below look like?

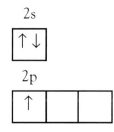

2s

↑ ↓

2p

↑ | |

65

Bond, Multiple Bond

Simple single-covalent bonds are also called *sigma* (σ) bonds. They are represented by a line running from the nucleus of one atom to the nucleus of another atom. A second kind of bond, called the *pi* (π) bond allows atoms to form multiple bonds and does so in a position such that the bonds do not run through the nuclei of the two atoms. The π bond is formed by the overlap of p orbitals.

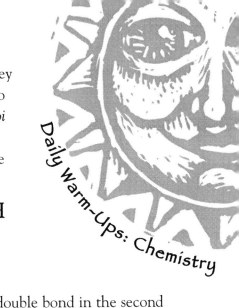

$$H-H \qquad \begin{array}{c} H \\ \end{array}\!\!\!\diagdown\!\!\!\!\!C\!\!=\!\!C\!\!\!\!\!\diagup\!\!\!\begin{array}{c} H \\ \end{array} \qquad H-C\equiv C-H$$

The H-H bond in the first molecule is one σ bond. The double bond in the second molecule is made of one σ and one π bond. The triple bond in the third molecule is made of one σ bond and two π bonds.

66

Draw a Lewis structure for N_2 and determine the number of σ and π bonds.

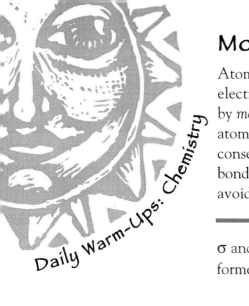

Molecular Orbital

Atoms have orbitals that are thought of as an area in an atom where electrons are likely to be found. Some forms of bonding are explained by *molecular orbital theory*, which is the idea that molecules, like atoms, have locations where atoms are most likely to be found and consequently have allowed locations where bonds may form. These bonds may be thought of as going from nucleus to nucleus, or as avoiding the area where the nucleus rests.

σ and π bonds are formed from molecular orbitals. If a triple bond is formed, where do the π bonds form?

67

© 2004 Walch Publishing

What's Its MO?

There are some guidelines to the formation of molecular orbitals (MO).

1. Each MO can hold, at most, two electrons.

2. MOs are formed most readily from atomic orbitals with similar energy levels.

3. The number of MOs formed is equal to the number of atomic orbitals combined.

___

If two lithium atoms, electron configuration $1s^2 2s^1$, were to form molecular orbitals, how would they form, and how many would form?

68

It's a Gas

One of the major states of matter is the gaseous state. Gases are characterized by several properties. They take the shape and volume of their container; unlike solids and liquids, they are easily compressed; and they readily form homogeneous mixtures with all other gases, regardless of chemical nature. The air we breathe is an example of a homogeneous mixture. Water's inability to mix with oil shows that this mixing is not as likely with liquids.

Why will a gas always expand to fill its container while a liquid will not?

69

© 2004 Walch Publishing

Under Pressure

Pressure is defined as the amount of force an object can exert on a certain area, or P = F/A. Common units for pressure are the Pascal (N/m^2) and the atmosphere (101.325 kPa, or 14.7 lbs/in^2). You may be familiar with pounds per square inch (lbs/in^2) from filling up a bicycle or car tire, but all of these are measures of pressure. Every gas exerts a pressure on any surface it contacts, which is caused by the collisions of the molecules in the gas with the surface it contacts.

How do the molecules of a gas exert pressure on the walls of a balloon?

70

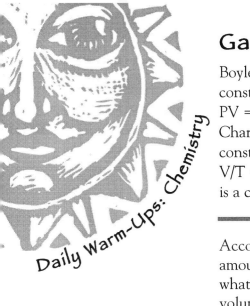

Gas Laws

Boyle's law states that the volume of a fixed amount of gas at a constant temperature is inversely proportional to the pressure, or PV = K where P is pressure, V is volume, and K is a constant. Charles's law states that the volume of a fixed amount of gas at a constant pressure is directly proportional to the temperature, or V/T = K where V is volume, T is the temperature in Kelvins, and K is a constant.

According to Boyle's law, what happens to the volume of a fixed amount of gas if the pressure is doubled? According to Charles's law, what happens to the temperature of a fixed amount of gas if the volume is cut in half?

71

Ideal Gases

The ideal gas law, PV = nRT, is an approximation that describes the behavior of gases quite accurately, especially when the pressures involved are low (near or less than two atmospheres). In the equation, P is pressure, V is volume, T is temperature in Kelvins, n is the number of moles of gas, and R is the ideal gas law constant 0.08206 L·atm/mol·K.

What is the volume of gas when 3.20 moles of oxygen are at 325 K and under a pressure of 1.13 atmospheres?

Partial to Gas

The pressure of a mixture of gases is equal to the sum of all the pressures of the individual gases in the mixture. This observation, called Dalton's law of partial pressures, states that the pressure each kind of gas exerts, called the partial pressure, can be added to all the other pressures to find the total. This is often represented by the formula

$$P_{total} = P_1 + P_2 + P_3 + \ldots$$

A mixture of three gases has a pressure of 3.34 atmospheres (atm). If the oxygen contributes a pressure of 0.75 atm and the nitrogen 1.13 atm, then what pressure does the helium gas contribute?

73

Kinetic Theory of Gases

The kinetic-molecular theory of gases states

1. Gases are made of many molecules that are in constant, random motion.

2. The sum volume of the molecules is negligible compared to the volume of the gas.

3. Intermolecular forces in a gas are negligible.

4. Energy can be transferred by collisions between molecules, but the average kinetic energy doesn't change, unless acted on by an outside energy source.

5. The average kinetic energy is proportional to the Kelvin temperature of the gas.

A sample of gas is compressed while its temperature is held constant. What happens to the velocity of the molecules in the gas?

74

Effusion and Diffusion

Effusion is a measure of the ability of a gas to escape through a tiny opening into an evacuated space—for example, air leaking through a pinhole in a spaceship. *Diffusion* is the ability of a gas to spread throughout a space, or another material, when its motion is essentially unrestricted, such as when the smell of perfume spreads from one side of a room to the other. Molecules of different masses diffuse and effuse at different rates. The lighter a molecule is, the faster it can diffuse or effuse.

Why would a lighter molecule diffuse or effuse faster than a heavier molecule at the same temperature?

Real Gases Aren't Ideal

Real gases obey the ideal gas law only under a few circumstances, such as when the pressure of the gas is extremely low. The reason these gases don't follow the ideal gas law very well is that there are assumptions made about the molecules in a gas that aren't true, but they make estimations easier. For example, real molecules are not infinitely small, and when they get close together, they attract or repel one another with great force.

Why would the volume of the molecules in the gas interfere with ideal gas law calculations?

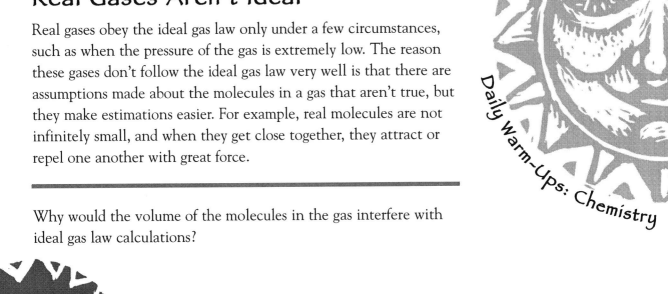

76

Sister States

There are many properties of liquids and solids that are similar. For example, they both have particles that are very close together, making them essentially incompressible. But they also have a few differences. The particles in a solid tend to remain in place, so the shape of a solid tends to change very little. Liquids, on the other hand, have a chaotic organization of particles that allows them to flow and take the shape of their container, something solids cannot do unless they are crushed into very fine pieces.

How would the rates of diffusion compare for a solid and a liquid?

77

© 2004 Walch Publishing

Hydrogen Bonding

In many molecules, hydrogen atoms that are connected on the periphery have the ability to exert their polar nature. In a polar bond, the hydrogen atom has a weak negative charge that can become attracted to a molecule with a weak positive charge, allowing the two to become connected by what is called a *hydrogen bond*. The most common example of this is water. It accounts for the high surface tension of water, which few other liquids of similar molecular mass and density have.

Which of the following molecules is likely to form hydrogen bonds?

a. H_2

b. H_2O

c. HF

d. CH_4

78

Surface Tension

Surface tension is a net inward pull of a liquid as the result of intermolecular forces. The beading of water on a smooth surface and the formation of nearly spherical drops of water on grass and other plants are examples of surface tension. The surface tension of water is about three times greater than that of most other liquids, a result of strong hydrogen bonds. The surface tension of mercury is almost six times greater than that of water.

How do you suppose surface tension is affected by temperature? Why?

79

Critical Issues

When a gas is placed under enough pressure, it will become liquid. At certain temperatures, however, the energy of the particles is so high that there is not enough attraction for the substance to be called a liquid. The highest temperature at which a substance can be seen in its liquid phase is called the *critical temperature*. Above this temperature no amount of pressure will compress the gas into a liquid. The pressure required to turn the gas into a liquid at exactly the critical temperature is called the *critical pressure*.

Does water have a high critical temperature or low critical temperature? Why?

80

Vapor Pressure

Vapor pressure is a measure of the pressure that is exerted, at a specific temperature, by the vapor from a solid or a liquid. Different substances have different vapor pressures, and they are often measured by placing a liquid in a closed container and then removing the air over the liquid. The molecules from the liquid will eventually turn to vapor, and the pressure increases until equilibrium is reached. At this point, a measure of the pressure at this given temperature is called the vapor pressure.

What problem would arise from trying to calculate the vapor pressure of hydrogen gas if the gas was collected by bubbling it through water?

81

Solid!

One of the things that separates solids from gases and liquids is the organized nature of their molecular structure. At the molecular level are crystals, which are repeating geometric shapes that are piled in a repeating manner into a crystal lattice. The crystal lattice helps give the solid defined planes and flat surfaces. Other kinds of solids, called *amorphous*, have a more random and chaotic internal structure that makes them more likely to lack definition and flat surfaces.

Why is a piece of iron considered a crystal even though it looks nothing like quartz or a diamond?

82

Solid Properties

The bonds in a solid have a lot to do with the properties of that solid. Some of the properties that bond types affect are melting point, boiling point, thermal and electrical conductivity, hardness, malleability, and ductility. Some of the types of bonds include covalent, ionic, metallic, and hydrogen bonds. Metallic bonds create substances that have a wide variety of properties, such as high thermal and electrical conductivity, malleability, and ductility.

What arrangement in metals accounts for their ability to conduct electricity?

83

LCD

As a solid melts, its organized internal structure becomes more chaotic and the crystal lattice destabilizes. Instead of passing from the solid phase to the liquid phase when heated, some liquids pass through an intermediate stage in which their internal structure has some properties of a liquid and some properties of a solid. Such substances are called *liquid crystals*, and they have more than one organized state to go through as their temperature changes.

Name three objects that have liquid crystal displays (LCDs).

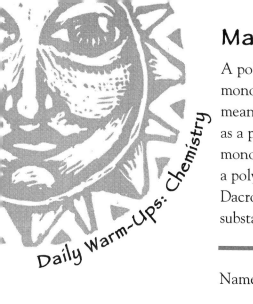

Many Units

A polymer is a substance made up of smaller repeating units called monomers. The word *polymer* is made up of two parts—*poly*, which means *many*, and *mer*, which means *unit*. You can think of a polymer as a paper-clip chain, in which each individual paper clip is a monomer, and hooking them together in a repeating manner makes a polymer. Many petroleum by-products are polymers, such as nylon, Dacron, rayon, and Styrofoam (polystyrene), as well as many natural substances, such as skin, rubber, and wool.

Name three uses for polymers in household items.

85

Biomedicine

Many polymers are formed to be used in biomedicine. These materials must be able to survive the harsh environment of the body and must be compatible with the body's other systems. Such devices replace heart valves, parts of joints, and blood vessels, to name a few uses. These materials allow humans to make their own replacement parts without waiting for organ donors and as alternatives to other expensive life-saving methods.

What are three concerns that must be addressed before a new biomaterial can be placed into a human body?

86

Ceramics

Ceramics are materials that can be crystalline or noncrystalline and are usually made out of inorganic and/or metallic components. Ceramics are usually brittle, hard, heat resistant, corrosion resistant, and wear resistant. Ceramics often replace metals in high-temperature applications, and they are typically much lighter than the metals they replace.

What is one disadvantage to using ceramics instead of metals?

87

Up, Up, and Conduct!

Superconductors are materials with little to no resistance to the flow of electricity. A copper wire is made up of many copper atoms, and as the electrons fight their way through the wire, there are many collisions and bumps that generate heat. One of the advantages of superconductors is that they require almost no energy to overcome the resistance found in a regular wire. One of their big disadvantages is that most superconductors only work at extremely low temperatures. Even the best superconductors found to date need to be in temperatures much lower than −100°C to work.

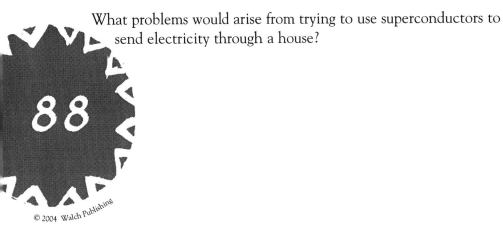

What problems would arise from trying to use superconductors to send electricity through a house?

88

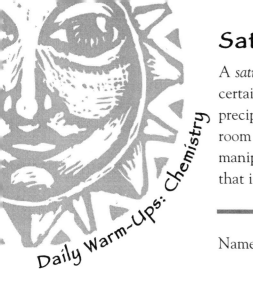

Saturation

A *saturated solution* is one that is holding all the solute it can at a certain temperature. The rate at which the solute dissolves and precipitates is at equilibrium. In an unsaturated solution there is still room for solute. In a supersaturated solution the solvent has been manipulated, usually by increasing and decreasing its temperature, so that it holds more solute than it usually can at that temperature.

Name a saturated solution that you have seen.

89

Solubility Traffic Jam

Some substances, such as oil and water, simply do not like to dissolve in each other, while other substances, such as alcohol and water, mix quite readily. Part of the reason for this is that the stronger the attraction between solvent and solute particles, the better the solubility. Water is a polar substance, oil is a nonpolar substance, and alcohol is a polar substance. Materials of the same nature, such as polar and polar, will dissolve each other. This is the basis of the chemists' phrase, "Like dissolves like."

Why doesn't oil dissolve in vinegar?

90

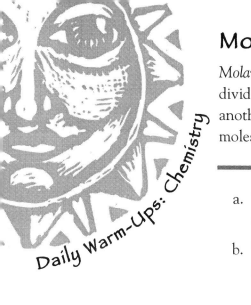

Molarity and Molality

Molarity (M) is a measure of concentration that is determined by dividing the moles of solute by the liters of solution. *Molality* (*m*) is another measure of concentration that is determined by dividing the moles of solute by the kilograms of solvent.

a. What is the molarity of a solution that has 3.4 moles of NaCl dissolved in 2.2 liters of solution?

b. What is the molality of a solution that has 13.4 moles of NaCl dissolved in 5.8 kilograms of water?

91

Colligative Properties

Colligative properties are those properties of solutions that depend only on the concentration of particles present and not on the type of particles. For example, the lowering of water's freezing point is a property that depends on the number of solute particles dissolved in the water and not the nature of the solute particles. The same is true for the boiling point elevation experienced by water when salt is added to it.

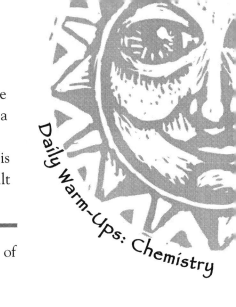

Which should have a greater effect in changing the boiling point of water when it dissolves? Why?

a. NaCl

b. $BaCl_2$

c. $C_{12}H_{22}O_{11}$

Colloids

A *colloid*, sometimes called a *colloidal suspension*, is a suspension of tiny particles between 1 nanometer and 1 micrometer in length. Because of their size, they do not settle out of a solution. The intermolecular forces and collisions that keep the particles suspended can be witnessed in milk, butter, cheese, fog, smoke, and even marshmallows.

Explain how whipped cream can be considered a colloid.

93

What Makes It Go?

The rate at which a chemical reaction takes place depends on many factors, but they can be summarized briefly:

1. the concentration of reactants

2. the temperature where the reaction is taking place

3. the state each of the reactants is in

4. whether there is a catalyst present

How would the state of a material affect its ability to react in a chemical reaction?

94

Reaction Rates

A *rate* is the measure of the amount of change that takes place in a certain amount of time. The rate of a reaction is a measure of how the concentration of reactants and products takes place over time. A common measure of concentration is molarity (M), and M/second is a common way of measuring reaction rates. The reaction rate describes either the rate at which a reactant is used or the rate at which a product is formed.

How would temperature affect a reaction rate?

95

Same or Different

A homogeneous reaction is one in which all of the reactants are in the same phase. One such reaction would be the reaction of hydrogen gas and oxygen gas to form water. A heterogeneous reaction is one in which the reactants are in more than one phase, such as when oxygen gas reacts with solid iron to form rust.

How would the state of a reactant affect how it interacts with other reactants?

96

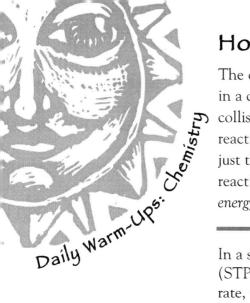

Hot Rates

The collision model is based on the idea that the particles of reactants in a chemical reaction must collide to react. If the number of collisions per second can be increased, then the rate at which reactions take place will also increase. The particles must collide at just the right angle, and they must have enough energy for the reaction to take place. This amount of energy is called the *activation energy*, and if it is not enough, then no reaction takes place.

In a simple mixture of gases at standard temperature and pressure (STP) there are more than 10 billion collisions per second. At this rate, why isn't every simple reaction over in a few seconds?

97

© 2004 Walch Publishing

Reaction Mechanics

Although a balanced chemical equation shows us the reactants and products to a reaction, it does very little to tell us about all the steps that might take place during the reaction. For example, the balanced equation

$$2NO \quad + \quad 2H_2 \quad \rightarrow \quad N_2 \quad + \quad 2H_2O$$

leaves out the intermediate steps and the rates at which they happen. For example:

$$2NO \quad \rightarrow \quad N_2O_2 \qquad (fast)$$

$$N_2O_2 \quad + \quad H_2 \quad \rightarrow \quad N_2O \quad + \quad H_2O \qquad (slow)$$

$$N_2O \quad + \quad H_2 \quad \rightarrow \quad N2 \quad + \quad H_2O \qquad (fast)$$

98

Why is it important for chemists to know all the intermediate steps in a reaction?

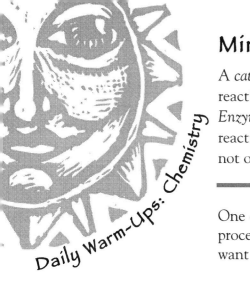

Mind if I Watch?

A *catalyst* is a substance that can increase or decrease the speed of a reaction without being consumed during the course of the reaction. *Enzymes* are one kind of catalyst that help speed up chemical reactions in the human body. Without enzymes the reactions might not occur fast enough to keep us alive.

One category of catalyst is an inhibitor, which slows some chemical processes. Give at least one example of a chemical process you might want to slow down.

99

All Things Being Equal

Chemical equilibrium occurs when a chemical reaction can proceed both forward and backward, and the opposing reactions take place at equal rates. It's similar to placing a half-full water bottle in the sun with the cover on. Some of the water will evaporate and then condense on the sides of the container, but after a certain point the rates of evaporation and condensation will be the same. Of course, in the case of water evaporating and condensing, no new products are formed. The formation of ammonia, shown below, is an example of chemical equilibrium.

$$N_2 \;+\; 3H_2 \;\rightleftarrows\; 2NH_3$$

By shifting temperature and pressure, chemists can control whether there are more reactants or products in a reaction that is in chemical equilibrium. Explain how this could be advantageous.

K_{eq}

The *equilibrium constant*, K_{eq}, is an expression of the ratio between product concentrations and reactant concentrations in a chemical reaction. It expresses the extent to which the equilibrium has proceeded in either direction of a reaction. When the value of K_{eq} is much greater than 1, this indicates that the products are favored, and equilibrium lies to the right of the reaction equation. When K_{eq} is much less than 1, reactants are favored, and equilibrium lies to the left.

What does it mean when the value of K_{eq} equals 1?

101

Calculate a Constant

For the reaction

aA + bB → cC + dD

$K_{eq} = [C]^c[D]^d/[A]^a[B]^b$

The capital letters indicate concentrations, and the lowercase letters are the coefficients from the balanced chemical equation.

In the following reaction, the concentration of hydrogen and iodine is 2.13 moles per liter (mol/l) for both, and the concentration of HI is 14.43 mol/l. What is K_{eq}?

H_2 + I_2 → 2HI

102

Predicting the Predictable

One method of predicting how a reaction will proceed is by determining the reaction quotient. The reaction quotient is calculated in the same expression as K_{eq}; however, the numbers put in the formula are from the initial conditions of the reactions. The reaction quotient, Q, is then compared to the known value of K_{eq}. If Q is greater than K_{eq}, the reaction will provide more reactants. If Q is less than K_{eq} the reaction will form more products.

What do we know about a chemical reaction where Q equals K_{eq}?

103

The More Things Change ...

Le Châtlier's principle states that if the equilibrium of a system is disturbed by a change in concentration, pressure, or temperature, the system will adjust itself to return to equilibrium. The disturbances can be described as either increases or decreases in concentration, pressure, or temperature. By observing the way these disturbances change the equilibrium of a reaction, we can see how such changes should affect any reaction. For example, increasing the temperature of an endothermic reaction increases the K_{eq} for the reaction.

What happens to the equilibrium when you decrease the temperature of an exothermic reaction?

104

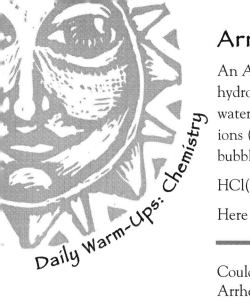

Arrhenius Acids and Bases

An *Arrhenius acid* is a compound that contains hydrogen and releases hydrogen ions (H^+), which then form hydronium ions (H_3O^+) in water. An *Arrhenius base* is a compound that can produce hydroxide ions (OH^-) in water. For example, when hydrogen chloride gas is bubbled through water the following process takes place:

$$HCl(g) \quad \rightarrow \quad H^+(aq) \quad + \quad Cl^-(aq)$$

Here $HCl(g)$ is the Arrhenius acid.

Could any of the following compounds be an Arrhenius acid?

    a. SO_2

    b. CH_4

    c. H_2S

    d. $Ca_3(AsO_4)_2$

105

Brønsted–Lowry

A *Brønsted-Lowry acid* is thought of as an H^+ donor. The H^+ ion is actually just a proton, so, by definition, a Brønsted-Lowry acid is a proton donor. The *Brønsted-Lowry base* is a proton acceptor. In the reaction below, the hydrogen carbonate ion acts as a Brønsted-Lowry acid by donating a proton, and water is the Brønsted-Lowry base because it accepts the proton.

$$HCO_3^- \ + \ H_2O \ \rightarrow \ H_3O^+ \ + \ CO_3^{2-}$$

Identify the Brønsted-Lowry acid and the Brønsted-Lowry base in this reaction.

$$HCl \ + \ H_2O \ \rightarrow \ H_3O^+ \ + \ Cl^-$$

106

Daily Warm-Ups: Chemistry

K_w

Water itself tends to be self-ionizing. In other words, even just a sample of pure water has the ability to form the common ions found in acid and base reactions. Although the amount of ionization in pure water is small, it exists and is represented by the formula below.

$$H_2O \quad \rightleftharpoons \quad H^+ \quad + \quad OH^-$$

The left-pointing arrow indicates that the reaction tends to favor a much larger quantity of water than of separate ions. Applying the law of chemical equilibrium, we get

$$\frac{[H^+][OH^-]}{[H_2O]} = K_w = 1.0 \times 10^{-14}$$

What is the value of $[H^+]$ in a neutral solution?

107

© 2004 Walch Publishing

pH

The current practice in chemistry is to evaluate the acidic or basic nature of a substance by calculating the pH of a solution. pH is found by taking the negative log, to the base 10, of the concentration of hydrogen ion.

$pH = -\log [H^+]$ and is sometimes given as $pH = -\log [H_3O^+]$

Calculate the pH of solutions that have an H^+ concentration of

       a. 1.0×10^{-4} M

       b. 1.0×10^{-8} M

       c. 1.0×10^{-12} M

108

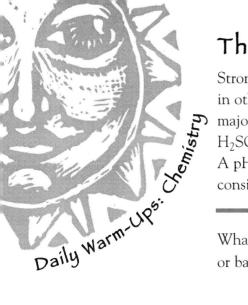

The Strong Acids

Strong acids and bases are substances that are also strong electrolytes; in other words, they exist in aqueous solution entirely as ions. The major strong acids are HBr, HCl, $HClO_3$, $HClO_4$, HNO_3, and H_2SO_4. The major strong bases are $Ca(OH)_2$, NaOH, and KOH. A pH from zero to 7 is considered acidic; pH values from 7 to 14 are considered basic. A pH of 7 is considered neutral.

What is the pH of a 0.042 solution of $HClO_3$? Is this solution acidic or basic?

109

© 2004 Walch Publishing

K$_a$

A *weak acid* is one that only partially ionizes in dilute aqueous solutions. K$_a$, the acid ionization constant, is the value of the equilibrium constant expression for the ionization of a weak acid. The weaker the acid, the lower the value of K$_a$, because these acids have the lowest concentration of ions and the highest concentration of nonionized acid molecules left over.

What does the value of K$_a$ tell us about the relative strength of a weak acid?

110

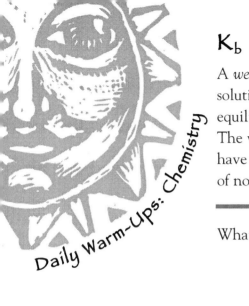

K_b

A *weak base* is one that only partially ionizes in dilute aqueous solutions. K_b, the base ionization constant, is the value of the equilibrium constant expression for the ionization of a weak base. The weaker the base, the lower the value of K_b, because these bases have the lowest concentration of ions and the highest concentration of nonionized base molecules left over.

What does K_b tell us about the relative strength of a weak base?

© 2004 Walch Publishing

K_a or K_b

As the value of K_a increases for a reaction, the value of K_b decreases, and vice versa. In a reaction with a conjugate acid-base pair, equilibrium is reached, and a K_a value and K_b value can be calculated for the same reaction. When the product of K_a and K_b is calculated, it equals 1.0×10^{-14}, which is the same as K_w.

---

a. Calculate K_a for the conjugate acid-base reaction of HF and F^- if $Kb = 1.5 \times 10^{-11}$.

b. Calculate K_b for the conjugate acid-base reaction of $NH4^+$ and NH_3 if $K_a = 5.6 \times 10^{-10}$.

112

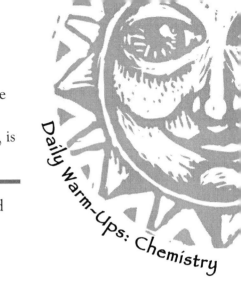

Aqueous Salts

Aqueous salt solutions that contain anions and cations that react with the water will cause a change in pH. A summary of these effects:

1. When both a conjugate acid and a conjugate base are present, the ion with the largest ionization constant will change the pH the most.

2. The conjugate acid of a weak base will cause a decrease in pH.

3. The conjugate base of a weak acid will cause an increase in pH.

4. The conjugate base of a strong acid won't affect the pH of a solution.

Why would an ion that is the conjugate base of a weak acid cause an increase in pH?

113

© 2004 Walch Publishing

Lewis

G. N. Lewis proposed a new method for classifying acids and bases that greatly increased the number of chemicals that could be considered acids. A *Lewis acid* is an electron-pair acceptor, and a *Lewis base* is an electron-pair donor. The point that allows more chemicals to be considered acids than in the Brønsted-Lowry acid definition is that Lewis established that the electron pair of the base could be donated to something other than an H^+ ion.

Identify the Lewis acid and Lewis base in the following reaction. (Hint: Draw the Lewis structures for each substance.)

$$NH_3 \quad + \quad H_2O \quad \rightarrow \quad NH_4^+ \quad + \quad OH^-$$

114

Common-Ion Effect

The *common-ion effect* is seen when more than one compound is dissolved into the same solution. One of the compounds must be a slightly soluble ionic compound and another must be a readily soluble ionic compound. The slightly soluble compound and the readily soluble compound must have one ion in common. The presence of the ions from the highly soluble compound makes it harder for the slightly soluble compound to dissolve.

$CuI(s)$ is slightly soluble in water, and NaI is highly soluble in water. How would the addition of NaI to a solution of CuI affect the equilibrium of the CuI?

115

Buffers

A *buffer* is a mixture of chemicals that help a solution resist the change in pH that usually accompanies the addition of a small amount of acid or base to a solution. When added to a solution, the resulting mixture is called a buffered solution. Blood is one example of a common buffer. The normal pH of human blood is usually between 7.3 and 7.5. In fact, if the pH of blood drops below 6.9 or rises above 7.7, it is often fatal.

An acid buffer is often made of a weak acid and one of its soluble salts. How does this allow the buffer to work?

K_{sp}

The *solubility product constant*, K_{sp}, is a number calculated by multiplying the concentration of each ion in a saturated solution of a slightly soluble electrolyte, after raising each concentration to the power of the coefficient found in the dissociation expression. For example, for the reaction below:

$$Ca_3(PO_4)_2(s) \leftrightarrow 3Ca^{2+}(aq) + 2PO_4^{3-}(aq),$$

$$Ksp = [Ca^{2+}]^3 \times [PO4^{3-}]^2.$$

Write the dissociation expression and solubility product constant expression for $Mg(OH)_2(s)$.

117

Precipitation Quotient

The *reaction quotient*, Q, is a term we can compare to the solubility product constant, K_{sp}. When K_{sp} = Q, equilibrium exists, and precipitation and dissolution occur at the same rate. When K_{sp} is less than Q, precipitation occurs until K_{sp} = Q. When K_{sp} is greater than Q, the solid continues to dissolve until K_{sp} = Q.

Will a precipitate form for a reaction where Q = 5.6×10^{-8} and K_{sp} = 2.6×10^{-11}? Why?

118

Quality versus Quantity

Qualitative analysis typically determines what substances a sample may contain. This is like judging the quality of a cake. Is it chocolate? Is there frosting? Is it made with extra-fattening ingredients, or is it sugar- and fat-free? *Quantitative analysis* is more concerned with the amount of each substance that can be found in a sample. We might wonder if there were two eggs or three in the cake. Was it two cups of flour or two and a half? How many grams of confectioners' sugar went into the frosting? Quality versus quantity tells us the difference between what is in a sample, and how much of each thing there is.

Determine if each of the following is a quantitative or qualitative measurement.

a. The sandwich has ham in it.

b. There are four eggs in the carton.

c. I used 230 grams of NaOH.

d. I added water to the solution.

119

Take a Deep Breath

The layer of the atmosphere closest to the earth's surface is called the *troposphere*. The troposphere occupies the first 10 to 16 kilometers closest to the earth. A typical sample of this air contains 78.1% nitrogen, 20.9% oxygen, 0.93% argon, 0.04% carbon dioxide, and traces of gases like xenon, hydrogen, krypton, ammonia, and carbon monoxide.

Is nitrogen poisonous?

120

Daily Warm-Ups: Chemistry

Electrons up There

A process called *photoionization* takes place at the outer reaches of the atmosphere. Photoionization is a process by which a molecule absorbs radiation, and the addition of that energy to the molecule causes it to lose an electron. In the upper atmosphere, the energy for this process comes from high-energy photons from the high-energy ultraviolet part of the electromagnetic spectrum. Because these wavelengths are absorbed, very little, if any, of this ultraviolet radiation ever reaches the surface.

Why is the atmosphere's absorbtion of high-energy ultraviolet radiation beneficial to living organisms?

121

© 2004 Walch Publishing

Ozone Alert

Although a large amount of ultraviolet radiation is absorbed during photoionization reactions in the outer atmosphere, some radiation still penetrates to the stratosphere. This energy is absorbed by O_2 molecules that break apart into O molecules and then combine with O_2 molecules to form ozone, O_3. The equation $O(g) + O_2(g) \rightarrow O_3(g)$ represents this reaction. Ozone is unstable because it has excess energy, which it usually loses through collisions with other molecules. It can then absorb high-energy solar radiation that causes it to photodissociate back into O and O_2 particles, allowing the process to start over again.

What problem seems to have been caused by the interference of chlorofluorocarbons (CFCs) with the ozone formation process?

122

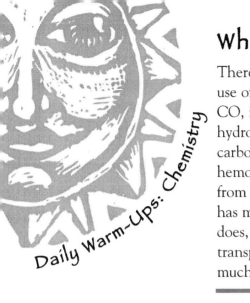

What's in a Breath?

There are some major pollutants that humans produce through the use of automobiles and various industrial processes. Carbon monoxide, CO, is a common product of the incomplete combustion of a hydrocarbon, perhaps seen most commonly in automobiles. When the carbon monoxide gets into a person's bloodstream, it is attracted to hemoglobin, the protein in red blood cells that helps transport oxygen from the lungs to places throughout the body. The carbon monoxide has more than 200 times the attraction to hemoglobin that oxygen does, and, as a consequence, it can prevent the hemoglobin from transporting oxygen, effectively suffocating a person who breathes too much of it.

How can we decrease the amount of CO produced by automobiles?

123

Salty!

More than 70% of the water on Earth is found in the oceans. This water has a salt content of about 3.5% by mass, which makes it unsuitable for drinking, watering crops, bathing, and a variety of other uses. Removing salt from water is a costly and time-consuming process, especially if attempted on a scale large enough to supply the water for a city or for large-scale agriculture. Distillation is one such process, where salt water is boiled—as the dissolved solids are left behind, the pure water vapor is condensed and collected.

What are some problems associated with distillation that would make it unwise to use for wide-scale desalinization of water?

124

Fresh!

Only about 0.6% of the water on Earth is considered fresh water. About 2% more is tied up in the glaciers and ice caps, but it is not readily usable. The water we have is in a fragile state because it comes into constant contact with humans and their ever-increasing need for fresh water. We use it to drink, cook, clean, bathe, wash vehicles, and dispose of waste, for agriculture, industrial production, and many other purposes.

Name three ways we damage our water supplies.

125

A Little Balance with Nature

A wide variety of chemists are trying to find a way to make chemistry as harmless as possible to the environment. Many large companies have been forced to start, or have voluntarily started, using methods that result in less pollution. For example, chemists are trying to use reactions that are energy efficient, produce little to no waste, use few extraneous chemicals, work with renewable chemicals, and produce by-products that might be used by other companies.

What are three ways that you might apply these simple guidelines to produce less waste in your own life?

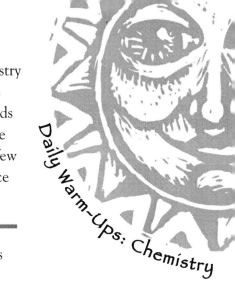

126

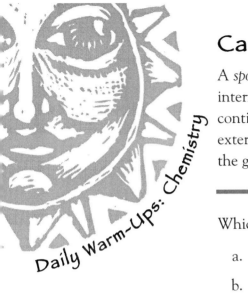

Can We Go Back?

A *spontaneous process* proceeds without any continuing outside interference. For example, a propane burner on a barbeque grill will continue to burn, after it has been lit, without the addition of any external energy. Likewise, a rock thrown off a cliff will fall until it hits the ground.

Which of the following reactions are spontaneous?

a. the condensation of steam to water at 95°C

b. A red-hot piece of metal is dropped into cold water and the water's temperature increases.

c. A crushed soda can pops back completely into its original shape.

127

What Entropy Means to Me

In general, *entropy* is a measure of the disorder in a system. For example, when the pins are set up in bowling, they are in a nicely organized pattern, with little entropy. As the bowling ball knocks down a few pins, the entropy increases. When all the pins are knocked down and scattered by a second ball, the entropy is very high. When the pins are set up for another frame, the entropy is low again.

Think of a situation in which the entropy increases and decreases in a repetitive manner.

128

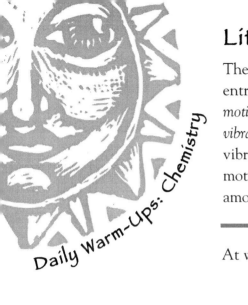

Little Entropy

There are three ways that molecules can move to increase their entropy. *Rotational motion* is the spinning of molecules; *translational motion* is the motion of the entire molecule off in a direction; and *vibrational motion* is the shaking of a molecule, much like the vibration of a guitar string back and forth in one location. As these motions increase, perhaps due to increased kinetic energy, the amount of entropy also increases.

At what point would such a molecule have an entropy of zero?

129

© 2004 Walch Publishing

Free Energy

The Gibbs free energy of a state is defined by the equation $G = H - TS$. The change in Gibbs free energy for a reaction at a constant temperature is $\Delta G = \Delta H - T\Delta S$ where ΔG is the change in Gibbs free energy, ΔH is the change in enthalpy, T is the absolute temperature, and ΔS is the change in entropy. When ΔG is positive, the reaction is not spontaneous forward, but backward. When ΔG is negative, the reaction is spontaneous in the forward direction, and when ΔG is zero, the reaction is in equilibrium forward and backward.

What can be said about the free energy in any spontaneous process at constant temperature and pressure?

130

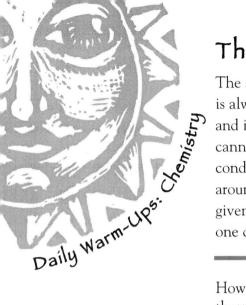

The Second Law

The second law of thermodynamics states that a spontaneous process is always accompanied by an increase in the total entropy of a system and its surroundings. If a process is spontaneous in one direction, it cannot be spontaneous in the other direction under the same conditions. This is an observation about the way we view the world around us and the everyday experiences that tend to show that for a given set of circumstances, processes tend to be spontaneous in only one direction.

How does the melting of ice at 2°C support the second law of thermodynamics?

131

On the OIL RIG

An oxidizing agent makes it possible for another substance to be oxidized. Such an agent is sometimes called an *oxidant* or an *oxidizer*. A *reducing agent* is a substance that loses electrons and allows another substance to be reduced. Such an agent is sometimes called a *reductant* or a *reducer*. The phrase OIL RIG provides a guideline for oxidation/reduction. **O**xidation **I**s **L**osing (of electrons) and **R**eduction **I**s **G**aining (of electrons).

Identify the oxidizing and reducing agents in the following reaction.

$$2Al \quad + \quad 3Br_2 \quad \rightarrow \quad 2AlBr_3$$

132

Balancing Redox Reactions

Redox reactions are quite often difficult to balance by inspection, or by trial-and-error. It is useful to understand that both the number of particles and the number of electrons must be balanced. In other words, the number of electrons lost by a substance that is oxidized must be gained by the substance that is reduced. For example, in the reaction

$$FeBr_3 \quad + \quad Zn \quad \rightarrow \quad ZnBr_2 \quad + \quad Fe,$$

iron undergoes a decrease of three in oxidation number, and zinc has an increase of two in oxidation number. This allows us to balance the equation:

$$2FeBr_3 \quad + \quad 2Zn \quad \rightarrow \quad 2ZnBr_2 \quad + \quad 3Fe$$

Balance the following redox reaction.

$$Zn \quad + \quad HNO_3 \quad \rightarrow \quad Zn(NO_3)_2 \quad + \quad NO_2 \quad + \quad H_2O$$

133

Anode and Cathode

A *voltaic cell* is an electrochemical cell that produces an electric current by means of chemical reaction in the cell that converts chemical potential energy into electrical energy. A battery is a common example of such a cell. The two poles of the battery are the cathode and the anode. On a battery, the negative terminal is the cathode, the electrode where reduction occurs. The positive terminal is the anode, the place where oxidation occurs.

During electrolysis of a substance such as water, what part of the electrolyte should be attracted to the anode?

134

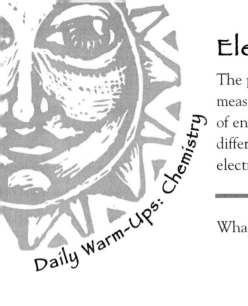

Electromotive Force

The potential difference between the two electrodes of a battery is measured in volts. A *volt* is the potential difference when one joule of energy is needed to move one coulomb of charge. This potential difference is called the *electromotive force* (emf) because it encourages electrons to move through the circuit.

What properties of a battery might affect its emf?

135

Did That Happen All by Itself?

Under nonstandard conditions, the electromotive force is represented by E. When the value of E is positive, the associated process must be spontaneous, and when the value of E is negative, the associated process must not be spontaneous. The standard cell potential is written as E° and is always written for reduction half reactions.

E° = E°(cathode) – E°(anode)

___

E° for the reduction half-reaction of $Cl_2(g) + 2e^- \rightarrow 2Cl^-(aq)$ is +1.36. Is the reaction spontaneous or not?

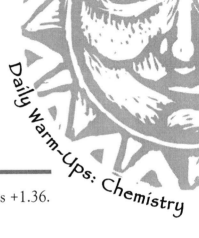

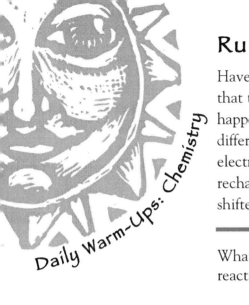

Running on Empty

Have you ever reached for a flashlight or a CD player only to find that the batteries were dead? From the chemical point of view, what happened was that the chemical reaction that supplies the potential difference reached equilibrium, and the ability of the batteries' electrons to be pushed or pulled ceased to have a net effect. In a rechargeable battery, the reaction can be reversed and the equilibrium shifted far to the left, allowing the net reaction to begin again.

What do we know about the value of E for a cell in which the reactants have reached equilibrium?

137

Portable Electricity

Battery is the common name for any collection of voltaic cells that are portable and provide a source of direct current. The typical battery is a collection of voltaic cells connected in series with the anode of one cell leading to the cathode of the next cell. When a collection of voltaic cells is connected in series, the total voltage of the battery is the sum of all the voltages on the voltaic cells. Dry-cell battery reactants cannot be recharged. In a wet cell the battery can be recharged, which allows the reaction to be reversed.

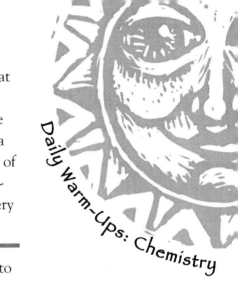

Six voltaic cells, each with a voltage of 1.5, are hooked in series to make a battery. What is the voltage on this battery?

138

Corrosion

Corrosion is an unwanted chemical reaction that damages a metal, usually through the formation of an unwanted product or by the dissolution of the metal. Corrosion reactions are typically spontaneous redox reactions in which a metal is oxidized, as in the reaction that turns iron (Fe) to rust ($Fe_2O_3 \cdot H_2O$). The water in the formula shows that when Fe^{2+} is oxidized to Fe^{3+} by dissolved oxygen in water, the insoluble substance rust is precipitated as a hydrate.

If a piece of iron is in water and the concentration of dissolved oxygen in the water is increased, will the amount of corrosion increase or decrease?

139

Electrolysis

In electrolysis an electric current is used to drive a reaction in a nonspontaneous direction. For example, in the reaction $2H_2(g) + O_2(g) \rightarrow 2H_2O(l)$, the spontaneous reaction is the formation of water from hydrogen and oxygen gas. In electrolysis the application of electricity forces the reverse reaction, as in the reaction $2H_2O(l) + \text{electricity} \rightarrow 2H_2(g) + O_2(g)$. Electrolysis of this manner is often performed by placing the cathode and anode of a cell in the same electrolyte.

At which electrode does oxidation take place during electrolysis?

140

Glowing in the Dark?

Radioactivity is the spontaneous breakdown of atomic nuclei that results in the release of some form of radiation, such as light, X rays, alpha particles, gamma rays, etc. Atoms that emit radiation from their nuclei are called radioisotopes, and the nuclei of these atoms are called radionuclides. A common particle emitted from such species is the helium-4 nuclei. The alpha decay of uranium is shown in this nuclear equation.

$$^{238}_{92}U \rightarrow ^{234}_{90}Th + ^{4}_{2}He$$

Notice that the mass numbers of 234 and 4 on the right of the equation add up to the mass number of 238 on the left, just as the atomic numbers of 90 and 2 on the right add up to the 92 on the left.

Write the nuclear equation for the alpha decay of Pu-239.

141

© 2004 Walch Publishing

Lead from Gold?

Just as alchemists dreamed of turning lead into gold, scientists dream of turning an element of one kind into an element of another kind. For example, by using nuclear transmutation, which is a process of bombarding atoms with particles to change their atomic numbers, scientists can take an element such as beryllium and turn it into carbon as in the following reaction. Note that there is a neutron left over, just as there may be other particles left over that must be accounted for.

$$\,_2^4\text{He} \quad + \quad \,_4^9\text{Be} \quad \rightarrow \quad \,_4^{12}\text{C} \quad + \quad \,_0^1\text{n}$$

What is the missing particle in the reaction below?

$$\,_7^{14}\text{N} \quad + \quad \,_2^4\text{He} \quad \rightarrow \quad ? \quad + \quad \,_1^1\text{H}$$

142

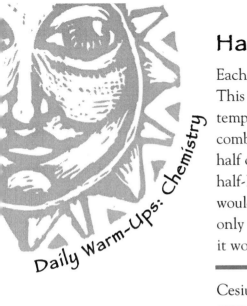

Half-life

Each kind of radioactive nucleus has a fixed rate at which it decays. This rate is unaffected by most normal external conditions, such as temperature, pressure, or whether or not they are chemically combined with another substance. The amount of time it takes for half of a sample of any such radioactive material to decay is called the half-life. For example, if you had a 100-gram sample of carbon-14, it would take 5,715 years for 50 grams of it to decay. After 11,430 years only 25 grams would remain, and after 17,145 years only 12.5 grams of it would remain.

Cesium-137 has a half-life of 30 years. What percent of a 100-gram sample would be left after 210 years?

143

Geiger Counter

A *Geiger counter* is a device designed to measure radiation. The Geiger counter detects the ionization that occurs within a low-pressure gas inside the Geiger counter. Just as a fluorescent lightbulb can be used to measure radiation leaking from a microwave oven, the Geiger counter's gas is ionized briefly by the passage of a radioactive particle, which allows a brief flow of electricity. The amount of electricity can be observed on a readout, or it can be converted into the familiar clicking sound that is often associated with a Geiger counter.

What does the radiation from a leaky microwave do to the gas in a fluorescent lightbulb that allows it to detect the microwave radiation?

144

Energy to Spare

When nuclear reactions take place, the corresponding energy changes can be very large. Einstein stated with the formula $E = mc^2$ that there should be a corresponding change in mass whenever there is a change in the amount of energy in a system. E is the amount of energy in joules, m is the mass in kilograms, and c is the speed of light, which is 3.00×10^8 m/s. Even with these large numbers, the amount of mass change is very small for most chemical process. For example, when 1,000 grams of water is cooled 100°C, it releases about 418 kJ of energy. The corresponding mass change is only about 4×10^{-10} grams.

How much energy is released when 1 gram of mass is changed to energy?

145

Gone Fission

Nuclear fission is a process whereby the nucleus of an atom is split into parts, and, as a result, a large amount of energy is released. Nuclear fission is how most of the nuclear reactors on Earth generate energy, which can then be turned into electrical energy. For example, in a fission reactor, uranium-235 is bombarded with neutrons to start the reaction seen below.

$$^{235}_{92}\text{U} \quad + \quad ^{1}_{0}\text{n} \quad \rightarrow \quad ^{141}_{56}\text{Ba} \quad + \quad ^{92}_{36}\text{Kr} \quad + \quad 3\,^{1}_{0}\text{n}$$

Note that three neutrons were generated that could go on to split other nuclei and create a chain reaction.

146

What is the missing component of the following nuclear fission equation?

$$^{235}_{92}\text{U} \quad + \quad ^{1}_{0}\text{n} \quad \rightarrow \quad ^{137}_{52}\text{Be} \quad + \quad ? \quad + \quad 2\,^{1}_{0}\text{n}$$

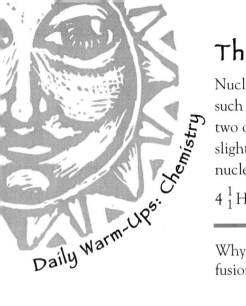

The Sun in a Bottle

Nuclear fusion typically takes place at extremely high temperatures, such as in the sun or other stars. In the process of nuclear fusion, two or more light nuclei combine to form one or more nuclei of slightly less mass than the combined mass of the original light nuclei. For example:

$$4 \, _1^1 H^+ \quad \rightarrow \quad _2^4 He \quad + \quad 2 \, _1^0 e^- \quad + \quad energy$$

Why must the kinetic energy of the reacting particles in a nuclear fusion reaction be extremely high?

Q: Can I Get in the Microwave? A: No.

There are three basic types of radiation that humans are exposed to. An *alpha particle*, which is essentially a helium nucleus, cannot penetrate the skin from the outside and can do very little damage. The *beta particle*, the electron, can penetrate the skin to a depth of about a centimeter. Both of these particles can do very little damage from the outside, but they can be harmful if they get inside the body. The *gamma ray* can penetrate any part of the body, and because of its extremely high energy, it can do a lot of damage.

Why would a source of alpha particles be dangerous if completely trapped inside of the body?

148

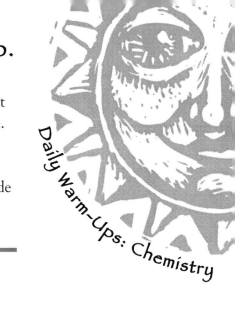

Daily Warm-Ups: Chemistry

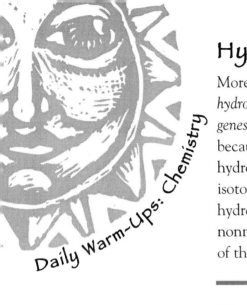

Hydrogen

More than 90% of the mass of the universe is hydrogen. The word *hydrogen* comes from the Greek roots *hydro*, which means water, and *genes*, which means to form. Hydrogen was named the water former because when it is burned in air, the oxygen combines with the hydrogen to form water. Hydrogen has three naturally occurring isotopes, hydrogen-1 (protium), hydrogen-2 (deuterium), and hydrogen-3 (tritium). Hydrogen exhibits properties of both metals and nonmetals, and it is the only nonmetal found in Group 1A, Column 1 of the periodic table.

What accounts for the fact that hydrogen can act like both a metal and nonmetal?

149

Boron

Boron is a metalloid found at the top of Group 3A, Column 13. It has little in common with the other elements in the same column. Boron is most commonly derived from the complex compound borax, which is found in deposits around the world, but most commonly in one large deposit in the Mojave Desert. Borax is used in fireproofing and cleaning, and one derivative called boric acid is used for disinfecting. Boron also forms compounds with hydrogen called boranes that are often used as additives in jet fuel.

Write the electron configuration for boron. Why might it be more likely to have a +3 oxidation number than other elements in the same column?

150

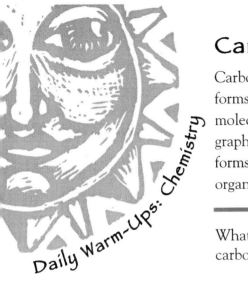

Carbon

Carbon is a very unusual element because it can exist in so many forms. Forms that have the same atoms, but are arranged in different molecular patterns, are called *allotropes*. Some allotropes of carbon are graphite, diamond, and fullerenes, also known as buckyballs. Carbon forms the backbone of almost every organic substance, both living organisms and nonliving sources such as petroleum.

What unusual bonding situation accounts for all the compounds that carbon can form?

151

© 2004 Walch Publishing

Carbon's Cousins

The elements of Group 4A, Column 14 (except for carbon) become more metallic in nature as one moves down the column. Silicon and germanium are metalloids, while tin and lead are metals. Silicon and germanium are both used in the construction of computer components, and silicon is the second most abundant element in the earth's crust. Lead and tin are very similar in that they are both soft and highly malleable; their major difference is that lead is highly toxic and is one of the three major heavy metals.

What is one thing that might account for the highly metallic nature of the elements at the bottom of Group 4A?

152

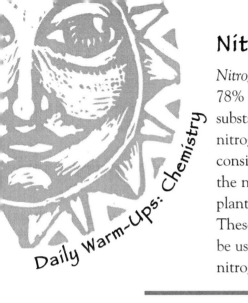

Nitrogen

Nitrogen is a colorless, odorless, and tasteless gas that makes up about 78% of the earth's atmosphere. Most living organisms have essential substances in them, like protein, that contain nitrogen and need other nitrogen-containing compounds to live. Interestingly, the atmosphere consists primarily of nitrogen, but few organisms can use it. Much of the nitrogen that does get into living creatures does so by way of plants that have bacteria around them in the soil and in their roots. These bacteria fix the molecular nitrogen into compounds that can be used by plants. The plants are then consumed and pass on the nitrogen-containing compounds.

How is it that we can breathe an atmosphere that is 78% nitrogen and not suffocate?

153

Nitrogen's Cousins

The elements of Group 5A, Column 15, other than nitrogen, consist of a nonmetal, phosphorus; two metalloids, arsenic and antimony; and a metal, bismuth. As one moves down Column 15, the metallic nature of the elements increases. Phosphorus occurs as three allotropes—red, white, and black. The white version spontaneously combusts in the presence of air. Arsenic is toxic, although once it was used in some medicines. Antimony has been used for darkening some forms of makeup. Bismuth has been used in some antidiarrheal medicines and is also used to make alloys with low melting points, such as those used in sprinkler systems.

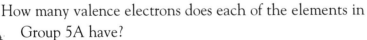

154

How many valence electrons does each of the elements in Group 5A have?

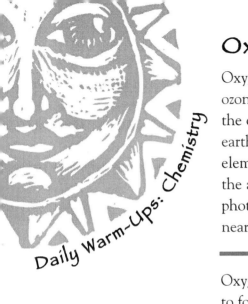

Oxygen

Oxygen has two allotropes—O_2, the version that we breathe, and O_3, ozone. The breathable form of oxygen, O_2, makes up about 21% of the earth's atmosphere. Oxygen is the most abundant element in the earth's crust, and it forms a compound with almost every other element except for the noble gases. A large portion of the oxygen in the atmosphere can be accounted for because it is a by-product of photosynthesis found in the rainforests, the algae in the oceans, and nearly all other plants.

Oxygen has six valence electrons. Why does this afford it the chance to form both ionic and covalent bonds?

155

Oxygen's Cousins

The elements of Group 6A, Column 16, other than oxygen, consist of two nonmetals, sulfur and selenium; and two metalloids, tellurium and polonium. Sulfur is used in the production of sulfuric acid, which is a component of fertilizers, steel, and paints. Selenium is used as a mineral supplement in vitamins, as well as in solar panels and in photocopiers. Tellurium is used in semiconductors, ceramics, and alloys. Polonium is radioactive and is used for reducing static charge, as a source of neutrons, and even as a power supply on the space shuttle.

Write the electron configuration of polonium.

156

The Halogens

The *halogens* are the elements in Group 7A, Column 17 of the periodic table. The word halogen means *salt formers*, and they consist of four nonmetals, fluorine, chlorine, bromine, iodine; and one metalloid, astatine. Astatine is radioactive, and due to the incredibly small amounts produced and the very short half-life of only about eight hours, there are practically no uses for it. The other halogens are reactive nonmetals that tend to gain or share one electron to form bonds, form –1 ions. They are essentially found combined in nature.

Write the electron configuration for iodine. What does this show about the valence electrons of the halogens?

157

Better Than the Rest

The noble gases are in Group 8A, Column 18 of the periodic table. The noble gases have a very stable arrangement of electrons, with eight in the outer level, filling the valence shell. They are all nonmetals, and because of their incredibly stable nature, they do not form compounds in nature, although compounds have been made with them under laboratory conditions. The first one discovered was helium, which was found in the emission spectrum of the sun.

How does the electron configuration of each noble gas end?

158

Through the Center

Groups 3B–2B, Columns 3–12 on the periodic table are the d-block elements, sometimes called the transition elements. This section of the table contains most of the metals that the average person thinks of as a metal in a nonchemical sense. Iron, tin, gold, silver, titanium, and copper are all part of this section of the periodic table, and they are indeed some of the most widely used metals on Earth. Many of the elements in this area lose their s-block valence electrons and sometimes their d-block electrons. This gives them variable valence states.

What properties might we expect from the transition elements?

159

Fire Makes It Better

Pyrometallurgy is the process of using high temperatures on mineral ores to separate the pure metal that is chemically combined in the ore. A fairly common way this is done is with copper.

$$2CuFeS_2(s) + 3O_2(g) + heat \rightarrow 2FeO(s) + SO_2(g) + 2CuS(s)$$

Then the CuS is heated in the presence of oxygen, and liquid copper is released in a process called smelting.

$$CuS(s) + O_2(g) + heat \rightarrow Cu(l) + SO_2(g)$$

Why is pyrometallurgy so important as rich deposits of certain metals become harder and harder to find?

160

Water Makes It Better

Hydrometallurgy is the process of removing minerals from their ores by a reaction that takes place in aqueous solution. Perhaps the most common of these processes is called *leaching*. In leaching, the compound that contains the metal that the refiner is interested in is dissolved by a solution that is specifically designed to single out just that compound.

What is one advantage that hydrometallurgy has over pyrometallurgy?

161

Electricity Makes It Better

Electrometallurgy is a process that uses electrolysis to reduce metal ores or to further refine metals into a purer form. Such processes can be carried out in either aqueous solution or in the molten salt that contains the desired metal. Aluminum, copper, and sodium are all metals that are obtained through this process.

Why would it be favorable to produce sodium metal through electrometallurgy instead of through hydrometallurgy from an aqueous solution of NaCl?

162

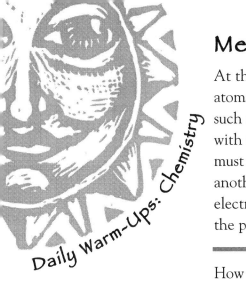

Metallic Bonding

At the atomic level, the crystals that are formed in many metals have atoms that are very closely packed. Many of the valence electrons in such a structure are unable to make bonds in the form of pair bonds with their neighbors. For the atoms to be held together, the electrons must be more mobile and allowed to move from one bonding area to another. This sort of arrangement is often referred to as the "sea of electrons." In this model, each electron is free to move around, while the positively charged metal ions stay essentially in one place.

How is a metallic bond different from a covalent or an ionic bond?

163

Alloys

An *alloy* is a combination of two or more metals. There are *homogeneous alloys*, such as brass and bronze, in which the two metals are spread evenly throughout the alloy. In a *heterogeneous alloy*, there are areas at the molecular level that can have fairly different structures, like some of the amalgams of mercury. Alloys tend to have properties that are a blend of the metals that are used to make them. For example, the steel that is used to make car and house keys is too hard to be cut with simple tools, so it is softened by adding lead to the alloy.

Although car and house keys are strong, shiny, and practical, why shouldn't they be given to babies to play with?

164

Magnetism

Magnetism is the ability of a substance to be affected by a magnetic field. *Diamagnetism* is when a substance is slightly repelled or totally unaffected by a magnetic field. *Paramagnetism* is when a substance is made magnetic by placing it in a magnetic field. The amount of magnetism is directly related to the strength of the magnetic field, and it dissipates when the magnetic field is removed. Substances that are strongly attracted to a magnetic field exhibit *ferromagnetism*. When substances like iron, nickel, and cobalt are placed in a magnetic field, their ions align themselves in the direction of the field and stay there, making them permanent magnets.

When an electric current is run through a metal nail, it becomes a temporary magnet. What kind of magnetic substance is it?

165

Complexes

A *metal complex* is a group of molecules or ions that are bonded to a central metal ion. A compound that contains such a complex is called a *coordination compound*. Many of the coordination compounds contain a transition metal as the central metal ion in the metal complex. One example of a metal complex is $Fe(CN)_6^{4-}$, where iron is the central metal ion with a charge of +2 and is surrounded by six cyanide ions in an octahedral arrangement.

Account for the charge of +2 on the complex $Fe(H_2O)_6^{2+}$.

166

Ligands

In a coordination compound the molecules or ions that surround the central metal ion are known as *ligands*. Just as ligaments are used to form connections inside the human body, ligands form connections to the metal ion with at least one lone pair of electrons. These electrons form a coordinate covalent bond to the central ion, so we say that the ligand coordinates the metal.

Draw the Lewis structure of water, and show where the electrons are that allow water to be a ligand.

167

Isomerism

Isomers are compounds that have the same chemical formula but different molecular structures. There are two major kinds of isomers. The first is the *structural isomer*, which is characterized by atoms that are bonded in different orders even though they have the same formulas. Such compounds have different chemical and physical properties. The second kind is the *stereoisomer*, which is when all the atoms are bonded in the same order but have a different arrangement in space.

Identify whether the structures below are structural isomers or stereoisomers.

168

```
 H   H   H   H
  \ /     \ /
   C       C
  / \     / \
 H   C = C   H
     |       |
     H       H
```

```
 H   H
  \ /
   C              H
  / \            /
 H   C = C      H
     |     \   /
     H      C
           / \
          H   H
```

Organic Chemistry

Organic chemistry gets its name from a misunderstanding in science. It was thought that only living organisms could form organic compounds until German chemist Friedrich Wöhler made urea in 1828 from inorganic reactants. The term *biochemistry* is typically used now to describe the study of the chemistry of living organisms. All organic and biochemical compounds contain carbon as one of their main constituents, and most contain hydrogen, oxygen, or nitrogen as well. Certainly carbon's ability to form four covalent bonds has made it possible for there to be literally millions of organic compounds.

What is the difference between biochemistry and organic chemistry?

169

Basic Hydrocarbon Structure

Hydrocarbons are organic molecules that contain only hydrogen and carbon atoms, held together by covalent bonds. The bonds may be both *sigma* and *pi* bonds, which means that the carbon-carbon bonds can be single, double, or triple. Perhaps the simplest of all the hydrocarbons is methane, CH_4, which has a carbon at the center, surrounded by four hydrogens in a tetrahedral arrangement.

What would you expect about the stability of a triple carbon-carbon bond compared to the stability of a single carbon-carbon bond?

170

The Basic Organic Structures: Alkanes

Alkanes are hydrocarbons that contain only single bonds between the carbon atoms. Saturated hydrocarbons have only single bonds throughout their entire structures, so by definition all alkanes are saturated. Some common alkanes are methane, CH_4; propane, C_3H_8; butane, C_4H_{10}; and octane, C_8H_{18}, which is found in gasoline. The formula for an alkane generally follows the pattern C_nH_{2n+2}, where n is the number of carbon atoms.

Which of the following are saturated?

    a. CH_4

    b. C_2H_4

    c. C_2H_2

    d. C_2H_6

171

Cyclic Compounds

Alkanes can form in long, straight chains, or they may split into branched chains. They can also curl back upon themselves and form rings called *cycloalkanes*. Cycloalkanes generally have a formula of C_nH_{2n}, which makes sense because two of the bonds that usually hold hydrogens are used in this case to form a single bond that closes the loop.

Which of the following could be cycloalkanes?

      a. C_3H_8

      b. C_6H_{12}

      c. C_8H_{18}

      d. C_5H_{10}

172

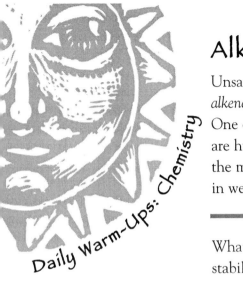

Alkenes and Alkynes

Unsaturated hydrocarbons contain double or triple bonds. The *alkenes* are hydrocarbons that contain a double carbon-carbon bond. One common alkene is polypropylene, $CH_3–CH=CH_2$. The *alkynes* are hydrocarbons that contain a triple carbon-carbon bond. Perhaps the most common of the alkynes is acetylene, C_2H_2, which is used in welding.

What could be said about the stability of acetylene compared to the stability of ethane, C_2H_6?

173

© 2004 Walch Publishing

Aromatic Compounds

In organic chemistry, aromatic compounds contain benzene as part of their structure. The word *aromatic* has long been associated with these compounds because many of the compounds came from the pleasant-smelling oils that were derived from fruits, flowers, and spices. When some other structure is attached to a benzene ring it is called a *substituted aromatic compound* and can exhibit properties quite different from that of benzene alone. A model of benzene and a substituted benzene ring appear below.

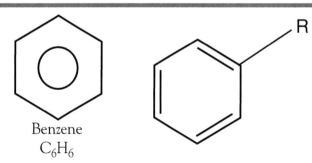

Benzene
C_6H_6

Benzene is a very stable molecule. Why does the addition of a substituted group greatly increase its reactivity?

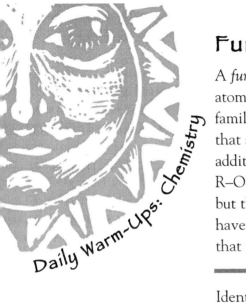

Functional Groups

A *functional group* in an organic compound is an atom or a group of atoms that typically behave in a certain way. Many of the different families of organic compounds are identified by the functional groups that are attached to them. The alcohol family is identified by the addition of a hydroxyl group, –OH, and has the general formula R–OH, where R represents any hydrocarbon. The ethers are similar, but they contain an oxygen that connects two hydrocarbons, and have the general formula R–O–R', where R' is a second hydrocarbon that may be the same as or different than R.

Identify the following as alcohols or ethers.

    a. CH_3OH

    b. C_2H_5OH

    c. $C_2H_5OC_2H_5$

175

Carbonyl Functional Groups

Carbonyl is a group of atoms that consists of a carbon double bonded to an oxygen. It makes up the backbone of four major organic families—the aldehydes, the ketones, the carboxylic acids (organic acids), and the esters. The general formulas for each are below.

Aldehydes

$$R-\overset{\overset{\displaystyle O}{\|}}{C}-H$$

Carboxylic acids

$$R-\overset{\overset{\displaystyle O}{\|}}{C}-OH$$

Ketones

$$R-\overset{\overset{\displaystyle O}{\|}}{C}-R$$

Esters

$$R-\overset{\overset{\displaystyle O}{\|}}{C}-O-R$$

Is there any special stability afforded to these molecules because of their structure?

176

Bio Comes Next

Biochemistry is the study of chemicals and chemical reactions that are inside living organisms. Many of the molecules studied in biochemistry are quite large and have to be built piece by piece from much smaller compounds found in the environment of living organisms. A lot of energy is needed to accomplish this feat, and the majority of the energy comes from the sun. This means that the organisms that have the ability to use energy directly from the sun, particularly in the form of photosynthesis, have an advantage over organisms that must acquire this energy in other ways.

How do animals get energy from the sun so that their bodies can create the large molecules often studied in biochemistry?

177

Muscle and Blood

Amino acids are organic molecules that have both an amino group and a carboxyl group. These amino acids combine to form proteins, which are organic polymers that carry out many jobs in living organisms. The body uses proteins to catalyze reactions, digest food, and form muscles, skin, fingernails, and cartilage. They are also used to transport oxygen as hemoglobin, regulate cellular processes, and recycle waste products in the body.

Given that proteins can be quite large, is it more efficient for the human body to make them from scratch or to consume them in ready-made form?

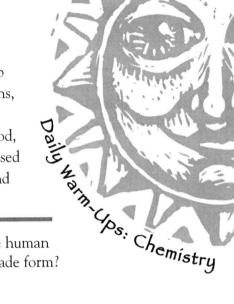

Carbo-loading

Carbohydrates are a group of organic substances that have the general formula of $C_n(H_2O)_n$. They are used as the major immediate energy source in organisms and can serve as a way of storing energy. Many people are familiar with blood sugar, or glucose, which has the formula $C_6H_{12}O_6$. This formula can be rewritten as $C_6(H_2O)_6$, and you can see that it fits the general formula. Many of these simple sugars can be derived from the breakdown of larger organic compounds, like disaccharides and polysaccharides.

What is the main function of carbohydrates in most living organisms?

179

Blueprint in the Middle

A *nucleic acid* is a biological polymer that contains nitrogen and is involved in the transmission and storage of genetic information. DNA is one such acid that contains the blueprint for building all of the proteins in a living organism's body. The monomers that make up the nucleic acids, called nucleotides, have three basic units:

1. a phosphate base in the form of phosphoric acid, H_3PO_4

2. deoxyribose, a five-carbon sugar

3. one of the four nitrogen bases—adenine, cytosine, guanine, or thymine, often represented by the letters A, C, G, and T

How might radiation affect the DNA of a cell?

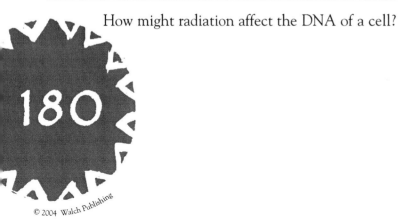

180

Answer Key

1. Answers will vary.
2. Any answer that contains a list of chemical and physical properties, such as color, odor, or flammability, will do.
3. Odor, flammability, density, boiling point, freezing point, etc.
4. The standard is reproducible anywhere on earth.
5. a. 3   d. 4
   b. 3   e. 2
   c. 1   f. 10
6. a. 1.64 years or 1 year and 235 days
   b. 1,400 years
   c. 14 quarters
   d. 9 meters
7. Answers will vary, but particles smaller than atoms have been discovered, and not all atoms of a single element are identical because of isotopes.
8. Proper experimentation keeps track of as many variables as possible, and good experiments are designed to be reproducible.
9. fluorine-19
10. Answers will vary, but medication is a common item measured in grams or milligrams. Items bought by mass would be the same amount whether purchased on Earth, on the moon, or in the weightless environment of free fall.
11. C and Si
12. HO, CH_2O, H_2O, CH_3
13. a. Li^{1+} d. Ca^{2+}
    b. 0   e. P^{3-}
    c. Br^{1-} f. Ga^{3+}
14. a. barium chloride
    b. potassium bromide
    c. lithium oxide
    d. magnesium oxide
    e. sodium sulfide
    f. beryllium nitride
15. a. dinitrogen dioxide
    b. carbon tetrachloride
    c. phosphorus hexabromide
    d. nitrogen monoxide
    e. tetrasulfur nonoxide
    f. pentacarbon heptiodide

16. a. $2C_2H_2 + 5O_2 \rightarrow 4CO_2 + 2H_2O$
    b. $2NaCl \rightarrow Cl_2 + 2Na$
    c. $3Ti + 2N_2 \rightarrow Ti_3N_4$
17. a. combination   c. combination
    b. decomposition   d. decomposition
18. a. 74.551 u   d. 94.108u
    b. 197.337 u   e. 56.078 u
    c. 283.896 u   f. 138.551 u
19. a. 6.71 moles
    b. 1.93×10^{24} atoms
    c. 29.1 moles
20. a. $CH_4 + 2O_2 \rightarrow CO_2 + 2H_2O$
    b. $2C_6H_6 + 15O_2 \rightarrow 12CO_2 + 6H_2O$
    c. $2C_{10}H_{22} + 31O_2 \rightarrow 20CO_2 + 22H_2O$
21. 40 molecules of hydrogen and 20 molecules of oxygen
22. hydrogen
23. As the amount of electrolytes in our bodies drops, the efficiency of the nervous system decreases.
24. a. $BaCl_2 + K_2SO_4 \rightarrow BaSO_4 + 2KCl$
    b. $NaOH + HCl \rightarrow NaCl + H_2O$
    c. $NH_4Cl + NaOH \rightarrow NH_4OH + NaCl$

25. a. acid   d. acid
    b. base   e. base
    c. acid   f. base
26. a. Magnesium is oxidized and oxygen is reduced
    b. carbon is oxidized, oxygen is reduced
27. a. 0.85 molar   b. 2.1 molar   c. 12 moles
28. five times, one half
29. Sound produces work by moving air particles and the mechanisms in sound receivers, such as microphones or ears.
30. 180 joules
31. a. positive   b. positive   c. negative
32. a. exothermic
    b. endothermic
    c. exothermic
33. 1,131.3 J
34. They are the same.
35. hydrogen and carbon
36. Hydrogen is still not mass-produced on a large scale, and for a large amount to be moved around, it must be under high pressure or extremely low temperature.

37. Blue light has a higher frequency than red.
38. 3.98×10^{-19}J
39. It would be very hard to keep track of all the subtle differences that make up the millions of substances that could be identified this way.
40. The object is too large to have its location or velocity changed by observation.
41. 1, 2, 3, 4, and 5
42. 32
43. no more than one electron in a box until they must double up

↑↓	↑	↑

44. $1s^2 2s^2 2p^6 3s^2 3p^6 4s^2 3d^{10} 4p^6 5s^2 4d^{10} 5p^6 6s^2 4f^{14} 5d^{10}$
45. a. f   d. p   g. s
    b. s   e. d   h. d
    c. p   f. f
46. Elements placed in order by atomic mass would be disorganized because of the existence of isotopes.
47. zero and two
48. a. Ca   c. Ge   e. W
    b. Y   d. Xe   f. Hg
49. Xe, Kr, Ar, Ne, He
50. The additional electron enters the s orbital in sodium and the p orbital in magnesium.

51. positive ions
52. negative ions
53. Hydrogen has no shielding electrons to interfere with the attraction between its only electron and the nucleus.
54. a. lose   d. lose
    b. gain   e. neither, already a noble gas
    c. gain   f. gain
55. a. lose 1   e. gain 3
    b. lose 2   f. gain 2
    c. lose 3   g. gain 1
    d. gain/lose 4   h. zero
56. Each oxygen atom would have to share two of its own electrons.
57. fluorine the highest, francium the lowest
58. Li•   •Ċ•   :F̈•

     •Be•   •N̈•   :N̈e:

     •B̈•   :Ö:

59. when the octet rule does not give an accurate picture of the actual molecule
60. Answers will vary, but one possibility is

:N̈:Ö̈:

61. The higher the value of bond enthalpy, the more stable the bond.
62. Some molecules with very small bond angles are slightly unstable due to repulsive forces between the crowded atoms.
63. The pair of nonbonding electrons in ammonia takes up a similar position as one of the hydrogens in methane.
64. Both chlorines have the same amount of pull, but chlorine and hydrogen have different amounts of pull.
65. The hybrid would be sp²

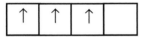

66. The molecule has one *sigma* bond and two *pi* bonds. :N≡N:
67. *Pi* bonds are located away from the nucleus and are usually parallel to one another.
68. Like energy levels would combine, and four MOs would form.
69. The intermolecular forces are much weaker in a gas than in a liquid, and the molecules can escape the attraction to their neighbors.
70. The molecules hit the balloon surface and apply a force to the surface. The force per unit of area is, by definition, pressure.
71. The volume is halved; the temperature is halved.
72. 75.5 L
73. 1.46 atm
74. If the temperature is held constant, the kinetic energy must also remain constant.
75. The lighter molecule has a higher velocity and will therefore spread faster or hit the opening to effuse more rapidly.
76. As pressure is increased, typically the volume is

decreased, and the percentage of the volume that the molecules take up increases.

77. Diffusion in liquids is slow but is far faster than it is in solids.

78. b and c; They are polar molecules with polar bonds.

79. As temperature increases surface tension decreases, because the increased velocity of particles decreases intermolecular forces.

80. Water has an unusually high critical temperature because of the high intermolecular attraction.

81. Some of the pressure would come from the water vapor.

82. The definition of a crystal is that it has a repeating geometric pattern at the molecular level.

83. The electrons in metals are highly mobile.

84. digital watches, laptop computer screens, hand-held video games, calculators, checkout register readouts, etc.

85. clothing, paint, any item made of plastic, etc.

86. Answers will vary. Some options are: How long before it wears out? Will the body accept or reject it? How much will it cost? Will it do the same job as the replaced part?

87. One disadvantage is that ceramics are brittle and tend to shatter instead of denting like metal.

88. It would be expensive to try to cool all the wiring in a house to $-100°C$ or colder.

89. any solution in which undissolved solute remains on the bottom, like sugar sitting at the bottom of a container of iced tea

90. Oil is nonpolar and water is polar.

91. a. $1.5 \, M$ b. $2.3 \, m$

92. b, because it will make the most particles per mole when dissolved

93. Whipped cream is a suspension of gas particles dispersed in a liquid.

94. The state of a material will affect the amount of surface area and the number of collisions that are possible with other reactants.

95. An increase in temperature generally increases the speed of the particles in a reaction, thereby increasing the rate at which they interact with one another.

96. Reactants in different states have different amounts of surface area and kinetic energy, which could speed or slow a chemical reaction.

97. There are two reasons: The number of particles could be much larger than 10 billion, and not very many collisions result in reaction because of improper angle of collision or not enough activation energy.

98. Knowing where the reactions slow down might allow chemists to add catalysts to speed the reaction up, or simply to control the overall rate better.

99. Answers will vary, but slowing down the rotting of food would have many benefits.

100. By controlling which side of the reaction is in excess, the chemist can actually make either side the product.

101. Neither reactants nor products are favored.

102. $[14.43]^2/[2.13][2.13] = 45.9$

103. The reaction's initial conditions are already at equilibrium.

104. K_{eq} increases, shifting equilibrium to the right.

105. Both b and c could be Arrhenius acids because they contain hydrogen.

106. The Brønsted-Lowry acid is HCl and the Brønsted-Lowry base is H_2O.

107. 1.0×10^{-7} M

108. a. 4   b. 8   c. 12

109. 1.38

110. The larger the value of K_a, the stronger the acid is.

111. The larger the value of K_b, the stronger the base is.

112. a. 6.7×10^{-4} b. 1.8×10^{-5}

113. It increases the basic nature of the solution.

114. The Lewis base is NH_3, and the Lewis acid is H^+ from the H_2O.

115. Equilibrium would be shifted to the left.

116. The buffer makes an equilibrium situation that only allows for small changes in the H^+ concentration.

117. $Mg(OH)_2(s) \leftrightarrow Mg^{2+}(aq) + 2OH^-(aq)$
$K_{sp} = [Mg] \times [OH]^2$

118. Yes, because K_{sp} is less than Q.

119. a. qualitative   c. quantitative
b. quantitative   d. qualitative

120. More than 70% of the air we breathe is nitrogen, so it is not poisonous to humans, but we need a fair amount of oxygen mixed with it to live.
121. The high-energy ultraviolet radiation can damage living organisms.
122. When CFCs interfere with the ozone process we are in danger of being exposed to more radiation from space, and the formation of the hole in the ozone layer is quite likely related to the use of CFCs.
123. by using fuels that don't produce CO as a by-product, or by changing the way automobile fuel is burned so that it undergoes complete combustion
124. expensive, time consuming, energy costs, disposal of left-over solids
125. Answers will vary, but some options are: pollution by fertilizers, pesticides, herbicides, and waste dumping.
126. Answers will vary.
127. a. spontaneous
     b. spontaneous
     c. not spontaneous

128. Answers will vary, but cleaning their rooms, playing a game with moving pieces, car getting dirty and then cleaned, are a few examples.
129. at 0 K
130. The free energy always increases.
131. The ice melts, and the particles of the water have less order than the particles of the ice.
132. Oxidizing agent = Br, reducing agent = Al
133. $Zn + 4HNO_3 \rightarrow Zn(NO_3)_2 + 2NO_2 + 2H_2O$
134. anions
135. electrolyte type and concentrations, composition of electrodes, temperature, etc.
136. The reaction is spontaneous.
137. The value of E is zero.
138. 9.0 volts
139. increase
140. the anode
141. $^{239}_{94}Pu \rightarrow ^{235}_{92}U + ^{4}_{2}He$
142. $^{17}_{8}O$
143. 0.78% or less than 1 gram

Daily Warm-Ups: Chemistry

144. The brief ionization caused by the radiation would cause flashes of light in the light bulb.

145. 9.0×10^{13} J

146. $^{97}_{40}$ Zr

147. They need to be moving fast enough to overcome the repulsive forces that keep them from combining.

148. Alpha particles couldn't escape from inside the body, so all of their energy would be transferred to the inside of the body.

149. Hydrogen can either gain or lose an electron to form either a negative or a positive ion.

150. The elements farther down the column are more likely to keep their s electrons.

151. Carbon has four sp^3 hybrid orbitals that allow it to form four covalent bonds.

152. The valence electrons are more loosely held.

153. Humans mostly metabolize the oxygen in the atmosphere, not the nitrogen.

154. five

155. Oxygen can easily gain two electrons or share two electrons.

156. $[Xe]6s^2 4f^{14} 5d^{10} 6p^4$

157. $[Kr]5s^2 4d^{10} 5p^5$, they all have seven valence electrons

158. with a full p orbital

159. high luster, electrical and thermal conductivity, malleability, ductility, and all other properties of metals

160. Pyrometallurgy allows us to get metals from any place where they are stored in ores, regardless of concentration.

161. Since hydrometallurgy doesn't require the high temperatures of pyrometallurgy, it doesn't require as much energy or create as much pollution.

162. The sodium produced would instantly be consumed by a reaction with the water.

163. The metallic bond holds a positive ion in place within a spread-out sea of electrons.

164. Car and house keys are alloys that contain lead, which is harmful to people and especially to young children or babies.

165. paramagnetic

166. The iron has a charge of +2, and all of the water molecules are neutral.

167. The structure looks as below, and either pair of unshared electrons could form the bond.

168. stereoisomers
169. Biochemistry is the study of chemistry in living organisms, while organic chemistry may include chemicals that come from nonliving sources.
170. The stability of the triple carbon-carbon bond should be much higher.
171. a and d
172. b and d
173. Acetylene's triple bond should make it more chemically stable.

174. The substituted group itself creates more reaction opportunities.
175. a. alcohol   b. alcohol   c. ether
176. Only the double bond between carbon and oxygen appreciates any special stability.
177. Animals eat plants or other organisms that ate plants to get the sun's energy, which is effectively stored in the plants and animals.
178. Consuming ready-made proteins (meat and other foods high in protein) is more energy efficient.
179. a source of immediate and stored energy
180. Radiation adds energy that can change the chemical makeup of the many substances that can be in each nucleotide.

Turn downtime into learning time!

Other books in the

Daily *Warm-Ups* series:

- Algebra
- Algebra II
- Analogies
- Biology
- Character Education
- Commonly Confused Words
- Critical Thinking
- Earth Science
- Geography
- Geometry
- Journal Writing
- Mythology
- Physics

- Poetry
- Pre-Algebra
- Prefixes, Suffixes, & Roots
- Shakespeare
- Spelling & Grammar
- Test-Prep Words
- U.S. History
- Vocabulary
- World Cultures
- World History
- World Religions
- Writing